TYRANNOSAURUS **SUE**

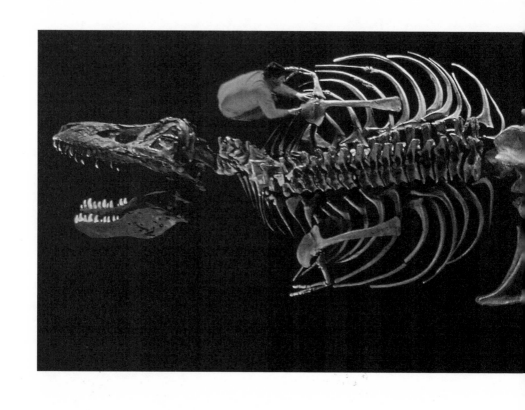

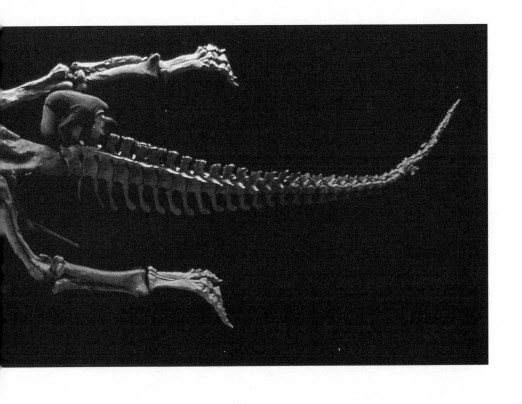

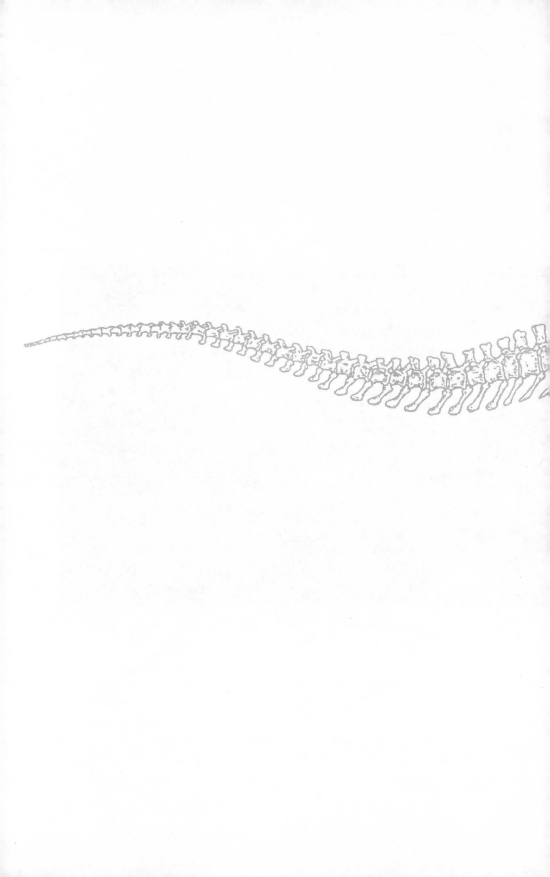

TYRANNOSAURUS SUE

THE EXTRAORDINARY SAGA OF THE LARGEST, MOST FOUGHT OVER *T. REX* EVER FOUND

STEVE FIFFER

FOREWORD BY ROBERT T. BAKKER

W. H. Freeman and Company
New York

For my family

Cover design by Stephanie Blumenthal
Text design by Nancy Singer Olaguera
Graphics by Dorothy Sigler Norton of Science Graphics, Bend, OR
Photographs: cover and pages ii–iii—Ira Block, NGS Image collection
pages 252–253—George Papadakis, © 2000 The Field Museum

Library of Congress Cataloging-in-Publication Data
Fiffer, Steve.
  Tyrannosaurus Sue : the extraordinary saga of the largest, most fought over
T. rex ever found / Steve Fiffer ; foreword by Robert T. Bakker.
    p.   cm.
  Includes index
  ISBN 0-7167-4017-6 (hardcover)
  ISBN 0-7167-9462-4 (paperback)
    1. Tyrannosaurus rex—South Dakota. 2. Paleontology—South Dakota—
History—20th century. 3. Larson, Peter. I. Title.

QE862.S3 F54 2000
567.912'9'09783—dc21   00-021596

Printed in the United States of America

First paperback printing 2001

W. H. Freeman and Company
41 Madison Avenue, New York, New York 10010
Houndmills, Basingstoke, RG21 6XS, England

Or speak to the earth, and let it teach you
Job 12:8

CONTENTS

FOREWORD

Robert T. Bakker, PhD

Dinosaurs are the most popular form of fossilized life the world over. And *Tyrannosaurus rex* is the most popular dinosaur among people of all ages, all cultures, and all nationalities. And the great skeleton called Sue is, by far, the most famous single skeleton of the "tyrant lizard king" ever found. A specimen of surpassing beauty, Sue can stun into silence even the most jaded and cynical senior scientist. Scholars everywhere agree that these bones are of tremendous importance in our understanding of the final days of the dinosaurs 66 million years ago, when the Age of Reptiles was about to end in worldwide catastrophe, and the legions of furry mammals were poised to take over the land ecosystems and evolve into horses and tigers, lemurs and monkeys, apes and mankind.

The story of Sue is an extraordinary tale—it has the romance of a paleontological quest, the search for the perfect *rex* by a young fossil hunter who, as a kid, dreamed of building a museum around a tyrannosaur specimen. The story has heartbreaking twists and turns, betrayals and palace intrigues, terrible moments of justice gone wrong when the best intentions were rewarded by calumny and imprisonment. Then there are unlikely protagonists: squads of fourth-grade children from Hill City, South Dakota, who tearfully pleaded with the FBI not to take their beloved tyrannosaur away; McDonald's and Disney, who rushed in their forces to save the skeleton from the ignominious fate of becoming the personal trinket of some international financier.

As the most perfect *T. rex* ever found, Sue belongs to the world. But she also belongs to Chicago, the city that hosts the Field Museum of Natural History, one of the oldest and grandest of America's fossil institutions. Here Sue, at last, has found a home, where she rises once more on her gigantic legs to thrill millions of visitors each year. And Sue's story most emphatically belongs to Steve Fiffer, a Chicago journalist who has followed her trail from the badlands of South Dakota through the federal court system and at last into the marble halls of the Field. Fiffer has done more than any other investigator in disentangling the many claims and counterclaims thrown out by dueling scientists and lawyers.

Fiffer's success comes from being thorough and being fair and being very smart. I know of no other investigator who has been welcomed equally by the rough-and-tumble folks who dig bones in the dry washes of the West and by PhDs in lab coats who speak in the polysyllabic language of technical paleontology.

There's a lot of Rashomon in Sue's story. Every participant remembers it from a different angle. One man's hero is another's villain. The supporters of Mr. Peter Larson, the independent fossil collector who dug Sue and made her famous, see him as a role model for anyone who wants to explore the deep past and bring back to life the great creatures of the Mesozoic. To certain segments of the PhDs in tax-supported university museums, Pete and his crew are Mesozoic brigands operating outside the tight circle of scholars trained in the Ivy League or its equivalent in California. When I visit my colleagues in Japan or Mongolia or Russia, I find Peter Larson regarded as a leading *rex* scientist who has reinvigorated the study of tyrannosaurs both by his discoveries of a half-dozen skeletons and by his bold hypotheses about Sue's family life and hunting style. But when I chat with long-time friends in Nebraska museums, professors of geology, the mere mention of the Larson brothers elicits sputtering curses and the most undignified slander.

Fiffer's account is a double story line. First, there is the chronicle of discovery, the ripping yarn of how packs of multiton *Tyrannosaurus rex* lived as top predators in the ecosystem known as the Lance Fauna, the last installment of Cretaceous history, and how this tyrant lizard king let its remains become entombed within the sandbars of ancient rivers flowing sluggishly out into the humid deltas of an inland sea. Buried with the tyrants' bones were a wealth of clues about what the *rex* smelled in the air and felt under their toes—there are beautiful tropical leaves, petrified fruit, and pollen dust that speak of a warm environment closer to present-day Louisiana than to the parched high plains of South Dakota. And there are remains of the tyrants' last meal, jagged-edged fragments of toes and ribs bitten off the bodies of *Triceratops* and still within the gut cavities of the predators' skeletons. Inscribed on the *rex* bones themselves are the scars left by a predaceous lifestyle—scratches on the cheekbones that were inflicted during ritualized battles with other *rex* and terrible wounds on legs and ribcages that were products of life-and-death struggles with horned dinosaurs.

The unearthing of these *rex* graves came in two waves. First came the explorers of 1900 from back East, from New York and from the Smithsonian, men who found the first *rex* and made "tyrannosaur" a household word. Nearly a century later, in the 1990s, came the second wave of *rex* hunters, this time led by men and women from institutions in the Rocky Mountain states, some from state colleges and universities, some from independent museums that grew de novo, coalescing around people such as the Larson brothers and their coworkers in the Black Hills of South Dakota. These independents were not a totally new phenomenon. Since the 1870s there had been Westerners—the Sternberg family from Kansas are the most famous—who made their living by digging dinosaurs and selling the specimens to public museums the world over. The Sternbergs braved horrible blizzards and scorching sandstorms to bring back duck-billed dinosaurs and tyrannosaurs to fill out the core displays in a dozen eastern museums.

When I went to college in the 1960s, the Sternberg legacy was still honored. But the politics of paleontology shifted. In the 1970s and 1980s some PhDs claimed all bones for themselves, declaring that only university scholars such as themselves had the right to dig skeletons and study them. According to this view, the days of the Sternbergs were over. Independent dino hunters were no longer welcome. Then came Sue. The Larson brothers' Black Hills Institute excavated the skeleton and began to clean the bones with expert care. And they dared to make their own conclusions about *rex* lifestyle and relationships. We PhDs are as vulnerable to the twin malady of envy and gossip as are any other segment of society, so as Sue's fame grew, resentment spread.

A byzantine tangle of lawsuits and, eventually, criminal charges paralyzed the further study of the Sue skeleton. Political opportunists jumped on the bandwagon, posturing as protectors of the country's fossil heritage. Men and women of goodwill lined up against each other. Former friends broke off communication.

When the dust settled, millions of taxpayer dollars had been squandered on unnecessary court cases. And Sue was in danger of being auctioned off to any billionaire collector who might want to make the fossil a Cretaceous objet d'art for his foyer.

Fiffer tells this double story with wit, clarity, and verve. Here's a fossil saga for everyone—for the dino aficionado who wants to learn secrets of rexian society; for the reader of adventure yarns that are almost too good to be true; for the citizens who suspect that the agents of big government don't always do the right thing. Dinosaur bones are the legacy given to our human species, the only life form with a brain big enough to understand them. Dinosaur fossils are intellectual jumper cables—nothing beats a room full of dino bones to get brains of kids and adults working at full capacity. And to understand dino science, one has to probe the motives of dino scientists. More than any other recent work, Fiffer's book gets inside paleontology and shows us how this most magical science really works.

ACKNOWLEDGMENTS

Reconstructing Sue's story was a little bit like reconstructing her skeleton. There were a lot of pieces that had to be put together, and the task was beyond the skills of any one person. Thus, I'm most grateful for the help of many.

Special thanks must go to the Black Hills Institute (Peter Larson, Neal Larson, Terry Wentz, and Marion Zenker) and the Field Museum (John McCarter, Amy Louis, John Flynn, Bill Simpson, and Nancy O'Shea). Sue Hendrickson, Bob Bakker, Phil Currie, and Ken Carpenter also provided essential material and guidance.

Many others with intimate knowledge of various aspects of the story graciously shared their thoughts with me, including Ed Able, James Abruzzo, Bob Chicoine, the late Gary Colbath, Peter Crane, Jack Daly, Patrick Duffy, Steve Emery, Henry Galiano, Richard Gray, Robert Hunt, Louis Jacobson, Bob Lamb, Robert Mandel, Amy Murray, Cathy Nemeth, David Redden, Keith Rigby, Dale Russell, Vincent Santucci, Kevin Schieffer, Maurice Williams, Michael Woodburne, and David Zuercher.

Several wonderful books were also extremely helpful in understanding the science of dinosaurs and the history of dinosaur hunting. As is obvious from the references in the text, I relied heavily on the following: *The Complete T. Rex* by John R. Horner and Don Lessem (Simon and Schuster, 1993); *The Rex Files* by Peter Larson (self-published); *The Dinosaur Hunters* by Robert Plate (McKay, 1964); *Dinosaur*

Hunters by David A. E. Spalding (Prima Publishing, 1993); *The Life of a Fossil Hunter* by Charles H. Sternberg (Indiana University Press, 1990; originally published by Holt, 1909); and *The Riddle of the Dinosaur* by John Noble Wilford (Knopf, 1985). "Jurassic Farce," a *South Dakota Law Review* article by Patrick Duffy and Lois Lofgren, was also a valuable resource. Other books consulted included *The Bonehunters' Revenge* by David Rains Wallace (Houghton Mifflin, 1999) and *Night Comes to the Cretaceous* by James Lawrence Powell (W. H. Freeman and Company, 1998). I have attempted to give credit in each instance in which I have directly quoted or used information from these texts. I apologize to the authors if I have failed to do so at any point in the book.

Newspaper articles also assisted me in reconstructing some events. Thanks to all the South Dakota publications that did such a fine job of covering this story, with special kudos to the *Rapid City Journal* and its writers Bill Harlan and Hugh O'Gara. The *Journal* provided outstanding coverage of this story from day one. Thanks, too, to Gemma Lockhart for her lovely commentary.

I would be remiss if I didn't acknowledge others who assisted: Richard Babcock, Ira Block, Diana Blume, Casey Carmody, Julia DeRosa, Georgia Lee Hadler, Karen Hendrickson, Penelope Hull, Maureen McNair, Stan Sacrison, Steven Sacrison, Minnie Tai, Jeff Theis, Judy Thompson, Sheldon Zenner, McDonald's, Walt Disney World Resort, Burson-Marsteller, Sotheby's, the staffs at *Chicago Magazine* and *Chicago Tribune Magazine*, and, as usual, my wonderfully supportive family— Sharon, Kate, Nora, and Rob Fiffer.

Penultimate thanks to my literary agent, Gail Hochman, and, at W. H. Freeman and Company, to Peter McGuigan and Sloane Lederer, who from the beginning saw the possibilities for this book, and to my editor, Erika Goldman, who provided invaluable support and counsel throughout the writing process.

And finally, thanks to Sue, herself, whose story and presence will, I hope, inspire a new generation of scientists and writers.

TYRANNOSAURUS SUE

PROLOGUE

After she was gone, after that dark, shocking day in May of 1992, when the armed FBI agents and Sheriff's officers and National Guardsmen had come and taken her away and locked her in a machine shop, he didn't forget her. When he wasn't talking to his lawyers about how to get her back or to the U.S. attorney who had ordered her seizure or to the media or to the leaders of the Cheyenne River Sioux tribe who also claimed her, when he wasn't talking to the technicians at NASA who'd been waiting for her or to the outraged scientists who couldn't believe she was gone, he'd climb into his 1981 Datsun pickup and drive the 30 miles from Hill City to Rapid City for a secret rendezvous.

Down Main Street and onto Highway 16, through the Black Hills past the turnoff for Mount Rushmore, past the entrance to Bear Country, the Reptile Gardens, the water slides, the miniature golf courses, and all the other attractions for the summer tourists, until he reached a brick building at the South Dakota School of Mines and Technology. The building, part of the school's physical plant, housed boilers and other heating systems, as well as the 40-foot metal storage container in which she was being held. He could catch a glimpse of the container through an outside window. "I'd just stand there and talk to her," Peter Larson remembers. "I'd say, 'We're gonna get you out of here, Sue. Be patient. Everything's going to be okay.'"

Larson, who was in his early forties, did not expect Sue, a 67-million-year-old *Tyrannosaurus rex*, to respond. Dinosaurs talk only in

1

movies or on television or in theme parks. Still, something in his bones told the paleontologist that he had to reassure Sue that she would survive this ordeal, this indignity, just as she had survived numerous battles with other dinosaurs while alive, just as her skeleton had survived millennia upon millennia of climactic change and chaos virtually intact.

Only 11 other *Tyrannosaurus rex* had ever been discovered. She was the find of his life, the find of *anyone's* life—the largest, most complete (90 percent) *T. rex* ever unearthed. And while she could not talk, she had been telling him remarkable stories over the 21 months since his colleague Sue Hendrickson had first spotted her on August 12, 1990.

Her bones offered clues to determining her sex and the sex of other dinosaurs, as well as the usefulness of their upper appendages. Her fibula, which had been broken and then healed over, seemed to indicate that she had survived a crippling injury that would have rendered her unable to fend for herself for a lengthy period of time. Her partially damaged skull indicated that she may have lost her life in combat. Foreign remains in her midsection even revealed her last supper before death— a duck-billed dinosaur. So to Larson, who had known he was going to be a paleontologist since he had found his first fossil at age 4, Sue was alive in her own way. She had a name, she had a personality, and she had a history. She was, to him at least, priceless.

As he took the podium at Sotheby's New York in midtown Manhattan on October 4, 1997, David Redden, the auction house's executive vice president, had no idea what price Lot 7045 would command. Nothing like this had ever been found, much less sold or auctioned, in the 253 years Sotheby's had been in business. But Redden, an unflappable Brit who has auctioned off everything from Mozart manuscripts to moon rocks, knew that the world would soon know the exact monetary value of the lot described in the Sotheby's catalog as a "highly important and virtually complete fossil skeleton of a *Tyrannosaurus rex* . . . popularly known as Sue." Bidding was to begin at $500,000.

Watching eagerly from a private room overlooking the standing-room-only crowd of 300 on the auction floor, Maurice Williams, the Native American who had consigned Sue to Sotheby's, was hoping for at least $1 million. Down on the floor itself, Sue Hendrickson prayed that it wouldn't get much higher than that. She knew that Peter Larson had a representative in the room with limited resources.

In the five and one-half years since the government had seized the dinosaur from him, Larson had exhausted his emotional and financial resources to get her back. His passion for Sue, he believed, had led to a vendetta by the federal government that put him behind bars for almost two years. Unable to attend the auction because he was under "home confinement" after his recent release from prison, the paleontologist was monitoring the proceedings via telephone.

A trio representing Chicago's Field Museum—Richard Gray, John W. McCarter, Jr., and Peter Crane—was also participating by phone, albeit one almost within shouting distance of Redden. Fearing that competitors' awareness of their interest in Sue might drive up the price, they had slipped unseen into another private room at Sotheby's just before the auction was to begin. They would communicate their bids by hotline to the house's president, Diana "Dede" Brooks, who was situated below the podium to Redden's left.

Less secretive than the Chicagoans but equally prepared for the hunt, representatives of several other museums as well as private foundations and wealthy individuals sat or stood on the floor, their weapons—small blue auction paddles—poised.

The United States government, which also had once laid claim to Sue, would not be bidding by paddle or phone. The government didn't care who in particular made the highest offer, but federal fingers were crossed hoping that the winner would keep the dinosaur in America. If some private collector or institution from Asia or Europe prevailed and took Sue overseas, Washington would be wiping the egg off its face for years to come.

"Good morning, ladies and gentlemen," Redden began. "We have for auction today the fossil of the *T. rex* known as Sue." The auctioneer tilted his head to the right, where Sue—or at least her chocolate-brown, 5-foot-long, 600-pound skull—sat on a long white cushion.

One of the world's foremost paleontologists, Dr. Robert Bakker, enamored by Sue's charisma, once described her as the "Marlene Dietrich of fossils." Sotheby's seemed to be treating her more like

Madonna than Marlene, a rock star rather than a chanteuse. Her image was posted on a huge screen above the stage, and a bodyguard stood by her side.

Redden took a deep breath, brought his gavel down, and the music began.

1

IT MUST BE A *T. REX*

"Five hundred thousand dollars," said Redden. He let the figure roll off his tongue slowly. The surreal nature of the moment hadn't escaped him. The bidding had begun, and he and a dinosaur named Sue had entered the kind of Salvador Dali landscape you might expect to find on the Sotheby's auction block.

If the tire hadn't gone flat, if the fog hadn't lifted, if the dinosaur hadn't been calling, Sue Hendrickson might never have found her. A *Tyrannosaurus rex* calling across 7 miles of rugged rock and barren prairie, across 67 million years of time? Hendrickson, of all people, can't explain it.

"I'm a very analytical person. I don't believe in fate," she says as she checks her mud-spattered gray backpack to make sure that she has everything she needs to go hunting for dinosaurs on a warm, cloudless, late-summer day in South Dakota. Her eyes as blue as the Great Plains sky, her face as resolute and weathered as the badlands she will soon be traversing, Hendrickson wears the same outfit she wore when she found Sue almost a decade earlier in this same fossil-rich formation—blue jeans, blue workshirt, and brown hiking boots. A silver-colored pick hangs from a leather belt around her waist like a six-shooter. Her hunting companion, a striking golden retriever named Skywalker, sits at attention by her side.

A field paleontologist since the mid-1980s, Hendrickson is nothing if not down to earth. It is past ages, not the New Age, that move her. She searches for fossils, not herself. And yet . . . .

"For two weeks this dinosaur was doing something to me, calling me. I didn't actually hear voices. But something kept pulling me there. Something wanted it to be *me* that went there and found her." (Paleontologist Robert Bakker is not surprised when he hears this. "There's no such thing as an atheist in a dinosaur quarry," he laughs.)

It was August of 1990. Hendrickson, Larson, and other members of a team from the Black Hills Institute of Geological Research, a commercial fossil-hunting enterprise based in Hill City, South Dakota, were concluding a month's stay at the Ruth Mason Dinosaur Quarry near the small town of Faith in the northwest corner of the state. This was the fourth summer that Hendrickson had worked with the team. She had joined them after a spring spent searching for amber in the Dominican Republic.

Paleontology is only one of Hendrickson's passions. She is, perhaps, the world's leading procurer of amber and has been an underwater diver on expeditions that have found the lost city of Alexandria, Egypt, submerged by an earthquake and tidal wave in the fifth century; sixteenth-century Spanish galleons that sank off Cuba and the Philippines; and Napoleonic ships sent to the floor of the Nile by Admiral Nelson's fleet in 1798. "Sue is like Indiana Jones, an intrepid globetrotter," says Dr. David Grimaldi, Chairman of the American Museum of Natural History's Department of Entomology.

Hendrickson was born with what she calls "itchy feet and the instinct to roam" in 1949 in—where else?—Indiana. A voracious reader, she devoured Dostoyevski by the time she was 11. Her mother, Mary, a retired school teacher, remembers her as a determined, intellectually curious girl, "who was too bright and too far ahead to fit in . . . a square peg in a round hole."

Hendrickson agrees that she had a hard time belonging after she reached high school in the middle-class town of Munster. At 16, she would tell her parents that she was going to a friend's house. Instead she'd make the half-hour drive across the state line to Chicago, where she'd listen to folk music in the city's Old Town section or sit out on Navy Pier and wish she were far away. "Typical teenage depression," she laughs. "I was bored. I hated my high school and I hated my hometown."

After several nasty clashes, mother and daughter agreed that a change of scenery was needed. Hendrickson moved in with her aunt and uncle in Ft. Lauderdale, Florida, for her senior year of high school. When they grounded her for staying out all night, she ran away with her boyfriend. "He liked to dive and I loved the water, so our plan was to work on a shrimp boat in Lafitte, Louisiana," Hendrickson says. On their first night there, however, a drunken local followed her out of a bar and tried to rape her.

Although she escaped the attack unharmed, Hendrickson had no desire to stay in the bayou country. She and her boyfriend took off, and for the next few years they crisscrossed the country, living and working odd jobs in cities from Boston to San Francisco, where, down to her last 36 cents, Hendrickson pawned her gold watch for $20. They finally decided to anchor themselves in the marina at San Rafael, California.

Hendrickson, who had called her parents immediately after she ran away and had visited them in Munster, asked her father to loan her the down payment on a 30-foot sailboat. He did. Soon she and her boyfriend were earning a living painting and varnishing the boats of their wealthy neighbors in the marina.

Two years later, the couple split up. Long fascinated by tropical fish, Hendrickson headed to Florida after learning that divers were often needed to catch museum and home aquarium specimens. She dove for a year in Key West, then moved to Seattle, where her parents had relocated.

Now 21, she contemplated college for the first time since running away. She passed the GED high school equivalency test and talked with the chairman of the Marine Biology Department at the University of Washington. "I asked him what I would get for seven years of study (needed to earn an undergraduate degree and PhD)," she recalls. "He said I could probably dissect fish or take pollution counts. I decided I could just go back and do what all the other PhDs were doing: catch tropical fish." After working as a sail maker for a year, she returned to Florida.

In 1973, while visiting diver friends in Key West, Hendrickson was asked to help salvage a sunken freighter. "It was very hard work," she says. "But I liked the challenge." Over the next few years, she helped raise sunken planes and boats. She earned additional money diving for lobsters and selling them to local restaurants. Her best haul: 492 lobsters in one day.

Hendrickson remembers that long before being pulled to the dinosaur, she had a knack for finding things. "I don't know if it's a sixth sense or luck, but there's something going on. When I'm attuned to something I'm looking for, I find it. Sometimes when I'd be looking for lobsters or seashells, I'd just stop my boat and say, 'This feels good.' I couldn't see a thing. I'd drop anchor in 30 feet, dive down, and there was the shell. It would blow me away. There was no sensory perception. I couldn't smell it or see it."

Hendrickson fell in love with the Dominican Republic while diving with a team of marine archaeologists there in 1974. She returned whenever she could. Always looking for new adventures, on one visit she went in search of the amber mines she had heard about from the locals. In the mountains, miners showed her their treasures—golden-hued lumps formed from prehistoric resins that imprisoned perfectly preserved scorpions, beetles, termites, and other insects. "It was amazing," says Hendrickson. "It was like seeing a whole other world, a window to the past."

On subsequent visits she tried digging in the mountain caves, but, she says, "It wasn't efficient. You could dig for years and never find an insect in amber." Instead, she began paying the miners for specimens. She didn't have to pay much at first—$10 to $35 per piece—because there was only minimal demand for such fossils.

At the time, most of the world's amber came from the Baltic Sea area (as it still does). Thus Hendrickson, who had become proficient in Spanish during her travels, was able to get in on the ground floor when she decided to supply specimens from Central America to museums and private collectors. In the mid-1980s, she began buying amber from miners in Chiapas, Mexico, as well.

Museums around the world study and display amber found by Hendrickson. "She has the nose for procuring pieces of big impact," says Dr. Robert K. Robbins, Curator of Entomology at the Smithsonian Institution. He cites one of the Smithsonian Insect Zoo's prize possessions, an extinct Antilles butterfly suspended in amber. Only six butterflies in amber have ever been found (three by Hendrickson), and, says Robbins, "This one is probably the best. You can even see the tiny hairs that served as its taste buds."

Hendrickson sells some amber pieces to private collectors, but she always offers her scientifically important finds to museums—at cost. Such generosity leads Bakker to observe, "Sue is the late twentieth-century equivalent of Alfred Russel Wallace." Like Wallace, the nineteenth-century British naturalist and explorer who independently reached the same theory of evolution as Darwin, Hendrickson seems motivated not by the desire for fame or fortune, but by a selfless "desire to share the earth's natural wonders," says Bakker.

Hendrickson describes herself in simpler terms. "I'm just a person who loves finding things and learning everything about them," she says, adding that she has little interest in possessing them. "It's the thrill of discovery. It's like the high from some drug. It lasts a few minutes. And it's addictive." She smiles. "Those moments are few and far between, but that's what keeps you going."

Hendrickson began looking for large fossils after meeting Swiss paleontologist Kirby Siber at a gem and mineral show in 1984. He invited her to help dig for prehistoric whales in Peru. There, in 1985, she met Larson.

The two soon became involved both professionally and romantically. On one of their first expeditions, they almost froze to death searching for meteorites in the Peruvian mountains. "We didn't have the right gear to stay up at that elevation overnight," Hendrickson says, "but we looked at each other and realized that this was too good an opportunity to pass up."

In 1987, Hendrickson began helping Larson hunt for dinosaurs and other ancient fossils at the Mason Quarry. She was, says Larson, a quick study. "Susan loves to learn," he says. "She gets attached to a subject and reads everything and asks questions. I'm the kind of person who will tell you ten times more than you want to know about something. Well, she wants to know that."

Larson had plenty to tell. He had been collecting fossils since 1956, when at the age of 4, he had spied a small brownish object in a ditch near the small farm on which he lived near Mission, South Dakota, some 200 miles southeast of Mount Rushmore. His parents took him to Mission, where friends June and Albert Zeitner, who ran a small geologic museum, identified the find as the tooth of an oreodont, a camel-like mammal that lived 25 million years ago.

"From that day on, I knew I wanted to hunt fossils," he says. "Here was something millions of years old, *extinct*, but it was still here. That was unbelievable. I loved living animals. I wanted to know the stories of the animals that weren't here anymore." Within a few years, he and his younger brother Neal and their older brother Mark had collected enough fossils and rocks to open up a "museum" in a 12-foot-by-15-foot outbuilding on their property, which lay within the borders of the Rosebud Indian Reservation. While other kids played cowboys and Indians, "we played curator," Larson says. They charged the adults in their family five cents admission.

By eighth grade, Larson had won the state science fair with an exhibit on fossils. "I was kind of a nerd," he confesses.

In 1970, he enrolled at the South Dakota School of Mines and Technology, one of the few schools he could afford to attend. There he majored in geology only because there was no major in paleontology. The pragmatic powers at the school pushed Larson and his fellow majors towards careers in the oil industry. Larson wasn't interested. "I wanted to hunt for fossils," he says.

Henry Fairfield Osborn, the distinguished scientist who founded the Department of Vertebrate Paleontology at the American Museum of Natural History in 1897 and then headed the institution when it rose to prominence at the beginning of the twentieth century, provided a job description of the fossil hunter, circa 1909:

> The fossil hunter must first of all be a scientific enthusiast. He must be willing to endure all kinds of hardships, to suffer cold in the early spring and the late autumn and early winter months, to suffer intense heat and the glare of the sun in summer months, and he must be prepared to drink alkali water, and in some regions to fight off the attack of the mosquito and other pests. He must be something of an engineer in order to handle large masses of stone and transport them over roadless wastes of desert to the nearest shipping point; he must have a delicate and skillful touch to preserve the least fragments of bone when fractured; he must be content with very plain living, because the profession is seldom if ever, remunerative, and he is almost invariably underpaid; he must find his chief reward and

stimulus in the sense of discovery and in the dispatching of specimens to museums which he has never seen for the benefit of a public which has little knowledge or appreciation of the self-sacrifices which the fossil hunter has made.

This was the ideal Larson aspired to when, after graduating from college in 1974, he and fellow student Jim Honert went into business— finding small fossils, rocks, and minerals and selling them to colleges. Their company, Black Hills Minerals, and its successor, the Black Hills Institute, weren't born of a desire to get rich. However, they were for-profit ventures. "I'm a capitalist," says Peter Larson. "I'm proud to be a capitalist. I think the capitalist system works. It creates money so that wonderful things can happen."

Larson and Honert quickly realized that colleges weren't the only institutions that wanted their finds. Says Larson, "It became increasing-ly obvious that museums really needed the service, because they had no way to get them. They didn't have the funds to send full-time people out searching for fossils or preparing fossils for exhibition."

The shift from hand specimens to museum pieces eventually sent Larson searching for the biggest fossil of all. In 1977, the Natural History Museum of Vienna, to which Larson had helped sell an ancient turtle, said it would like a dinosaur. By this time he was operating as Black Hills Minerals and had been joined by his brother Neal and Bob Farrar, both graduates of the School of Mines. "We said, 'No problem,'" Larson laughs, adding, "We didn't have a dinosaur. We didn't even know where to dig dinosaurs."

Two years later they journeyed to Faith, 150 miles northeast of Hill City, at the invitation of an octogenarian named Ruth Mason. Her land lies in a stretch of badlands called the Hell Creek Formation. As a young girl Ms. Mason had found what she thought were dinosaur bones on her property and had been trying to interest paleontologists in digging there since the early 1900s. The Larsons were the first to take her up on her offer.

In his book *The Complete T. Rex,* paleontologist Jack Horner, cura-tor of the Museum of the Rockies, based in Bozeman, Montana, pro-vides the following "recipe" for making a fossil: "An animal dies. Soon after death its flesh rots away. Over time sediment covers the bones.

That sediment compacts into rock. Minerals enter into the bone within the rock and preserve it." Recent experiments suggest that bacteria associated with the decaying carcass cause the minerals to precipitate out of groundwater, thereby fossilizing the bone.

Sand and silt are necessary recipe ingredients. Just add water—a slow-moving stream or river clogged with dirt or debris will do—and voilà! Of course, there is quite a bit of time between preparation and presentation; only after erosion occurs and the rock is swept away will the fossilized bones be visible.

The age of dinosaurs began during the Mesozoic Era about 225 million years ago and lasted for 160 million years, through the Triassic, Jurassic, and Cretaceous periods. The Hell Creek Formation and equivalent formations, which extend from South Dakota into Wyoming, Montana, and the Canadian province of Alberta, were the perfect kitchens for fossil creation during the final one or two million years of the Cretaceous period, which lasted from about 144 million BC to 65 million BC. What is now north central North America was warm and swampy then, with shallow seas and winding rivers—not cool and barren as it is today.

During the Cretaceous period the landscape was rich in flora similar to the vegetation one finds in the southern United States today—ferns, flowering plants, palm trees, and redwoods. Some of the creatures that moved about this landscape or flew above it or swam in the seas and rivers live on today. Birds filled the skies. Sharks and turtles swam the seas and rivers. Insects such as spiders and mammals similar to the opossum were abundant (although during the entire 150 million years dinosaurs ruled the earth, no mammal got much larger than a house cat).

Other Cretaceous creatures, which have not survived the last 65 million years, were also plentiful. Many of these were reptiles. The flying pterosaurs dominated the air. Fish-like ichthyosaurs inhabited the seas, as did giant marine lizards, the mosasaurs. And, of course, dinosaurs roamed the earth.

Many kinds of dinosaurs had already become extinct as the Cretaceous period drew to a close, but large herds of duck-bills and *Triceratops* still moved about. They spent their summers in the north, then migrated south to warmer climes for the winter. Dominating the landscape was another dinosaur—a ferocious carnivore that stood upwards of a dozen feet at the hip and upwards of 35 feet long from head to tail—*Tyrannosaurus rex.*

On Mason's land, the Larsons soon found remains of the duck-billed dinosaur, *Edmontosaurus annectens.* And not just one dinosaur. Inexplicably, thousands of duck-bills had died there and were deposited as a bone bed in the graveyard quarry.

The institute team spent the better part of the next two years excavating bones and then trying to piece them together to reconstruct a complete dinosaur. Finally, in the spring of 1981, Peter Larson took the assembled bones to Switzerland. There, he hoped to sell the specimen with the help of Kirby Siber, who had invested money in the project.

The two men had first met in the mid-1970s at a gem and mineral show in Tucson, Arizona. Larson, just starting out in the business, had brought several specimens to sell. He had mixed emotions about parting with his favorite, a pearl-white ammonite, an extinct relative of the chambered nautilus. "I marked it outrageously high, $700, because I really didn't want to sell it," he remembers. A French collector offered $450. "No thanks," said Larson. Finally, another buyer offered the full $700. This time Larson said, "Thanks." The institute could use the money.

Soon Larson learned that the purchaser, who turned out to be Siber, had immediately resold the ammonite . . . for $1400. Any ill will toward Siber was mitigated by the fact that the Swiss paleontologist bought several pieces from Larson at the show, giving the institute a bit of financial breathing room.

Although the duck-bill Larson took to Switzerland had been excavated over a three-year period, the institute had yet to receive any money for the thousands of man-hours already spent on the project at the Mason Quarry and in the preparation lab; Larson had never signed a contract with the Viennese museum. To stay afloat during this period, the institute had borrowed about $60,000, some at an interest rate exceeding 20 percent.

While in Switzerland, Larson learned that the museum did not have the funds to purchase the duck-bill. With no apparent means for meeting his loan obligation, he feared that he might have to go out of business. Fortunately, another commercial collector, Allen Graffham, put him in touch with the Ulster Museum in Belfast, Ireland. The museum paid $150,000 for the dinosaur, but Larson didn't see any of that money until 1988. After paying Graffham his sales commission and Siber for his investment, the institute ended up making less than $1 per hour on the transaction, Larson estimates.

Teams from the institute worked the quarry for duck-bills throughout the 1980s, eventually reconstructing nine more specimens, the last three bringing more than $300,000 each from museums in Japan, Europe, and the United States. During this period, Larson spent time collecting in South America as well as in South Dakota. In 1985 the institute and the Peruvian government entered into a partnership that yielded several scientifically important specimens, including a new family of sharks and a previously unknown marine sloth. Larson, Siber, and Hendrickson also donated their time and money to build a museum off the Pan American Highway south of the Peruvian city of Nasca. The museum features a 12-million-year-old baleen whale that is displayed where it was discovered—in the sands of a desert that was once ocean.

By 1990, the institute had become one of the largest suppliers of museum specimens in the world, doing business with, among others, the Smithsonian Institution, American Museum of Natural History, Field Museum, Carnegie Institute, Houston Museum of Natural Sciences, Denver Museum of Natural History, Natural History Museum of Los Angeles County, Yale's Peabody Museum, and museums in Germany, Japan, and Great Britain. The institute was also one of the largest employers in Hill City (population 650), with a full-time staff of eleven working out of the former American Legion Hall on Main Street. The bright white two-story Art Deco structure built by the WPA during the Depression housed the institute's offices, library, fossil preparation lab, storage area, and gift shop, and it featured a modest showroom that attracted a small percentage of the 2 million tourists who came to the Black Hills each year to visit nearby Mount Rushmore. This showroom had no *T. rex*. Rather, it exhibited considerably smaller finds such as a 7½-inch tooth from a 60-foot long prehistoric shark and numerous colorful ammonites.

Over the years, the institute has sold some finds for considerable amounts of money. Still, none of the principals has become wealthy. Excavation and reconstruction of specimens is extremely costly, and, as the case of the Ulster duck-bill illustrates, sales transactions sometimes take years. As self-described "Republican paleontologists," the Larson brothers rejected the idea of applying for government grants because of their distaste for bureaucracy.

Peter Larson, who lives in an old trailer a few yards from the institute's back door, cites another reason for his bare-bones existence. "We set aside the best specimens for the museum we'd always dreamed of building."

The Larson brothers did not charge admission to the institute's showroom, and they didn't plan on charging admission to the museum of their dreams, if and when it ever became a reality. "Education is the most important thing, and we don't believe people should have to pay for education," says Neal Larson. He and Peter and other institute staffers give 30 to 60 school talks a year, taking their fossil displays around the Black Hills area. They also speak to amateur groups, rock clubs, and colleges and take people out collecting free of charge.

The Larsons, Hendrickson, and an institute crew that included *1990* Peter's 10-year old son Matthew and Neal's 15-year-old son Jason, spent much of the summer of 1990 in the area around the Mason Quarry. Shortly after the fossil hunters arrived, they found a dead horse belonging to Maurice Williams, a one-quarter Native American whose large cattle ranch on the Cheyenne River Sioux Reservation lay just to the east of Ms. Mason's property. When Williams came by to claim the animal, he asked Peter Larson about the dig. "He was fascinated," says Larson. "He said, 'I've got land with badlands on it. Why don't you come over and look for dinosaurs?' I said, 'Great. We don't pay a lot, but if we find something of significance, we'll pay you.'" Williams also suggested that Larson call his brother Sharkey, who owned similar land in the area. Sharkey Williams, now deceased, also invited the institute to dig on his property.

While the institute team initially found little on Maurice Williams's land, they did find a few partial *Triceratops* skulls on his brother's property. On the morning of August 12, they were preparing to excavate a skull when they noticed a flat tire on their collecting truck, a rusting, green 1975 Suburban. They changed the tire and saw that the spare was also dangerously low on air. His tire pump broken, Larson decided to drive into Faith to get the two tires fixed. He invited Hendrickson to join him on the 45-minute drive.

She declined. The dig was to end in a couple of days, and she wanted to explore the sandstone cliff that had been calling her ever since she had spied it from several miles away two weeks earlier. "I'd kept thinking, I gotta get over there," Hendrickson recalls, "but you're so tired, just physically exhausted at the end of the day." Maurice Williams had asked that they keep their vehicles off his property, so Hendrickson knew that she would have to walk to the cliff over rugged terrain. Now she finally had the time to do it.

"It was foggy," she says. "It never gets foggy in South Dakota in the summer, but it was foggy that day." Although the rare mist prevented her from seeing her destination, Hendrickson set out with her dog on the 7-mile hike to the cliff. "I told myself, 'Don't walk in a circle,' but that's just what I did." After two hours she found herself back where she had started.

"I started to cry," she confesses. Her failure to find the cliff was not the only cause for tears. During their time at the quarry, she and Larson had mutually agreed to break up.

Determined to find out what had been calling her, Hendrickson waited until the fog lifted and began the trek a second time. Two hours later she stood at the foot of the 60-foot-high, buff-colored formation.

"It's not easy to find fossils," Larson says. "You're trying to figure out what's bone and what's not. You're seeing fragments and trying to imagine what they are, selecting which fragments are important. You have to have a good eye that can spot textural differences and color differences. Susan does."

Bakker adds: "There are some people who can find fossils and some who can't. Sue has the talent for getting a sense of place in a paleontological context. You must be born with it. She was. And she has honed it."

Hendrickson began her search by walking along the bottom of the cliff, eyes on the ground. "Usually you walk along the bottom to see if anything has dribbled down," she explains. "If you don't see anything, you walk along the middle of the formation, if it's not too steep. And then you might walk a third time across the top, just to hit the different levels."

Halfway through her first pass at the bottom, she saw a "bunch of dribbled-down bones." Where had they come from? Hendrickson looked up. "Just above eye level, about 8 feet high, there it was: three large dinosaur vertebrae and a femur weathering out of the cliff. It was so exciting because they were very large and because of the shape. The carnivores like *T. rex* had concave vertebrae from the disk; it dips in. The herbivores—the *Triceratops* or duck-bills is what you almost always find—have very straight vertebrae. So I knew it was a carnivore. I knew it was really big. And therefore I felt it must be *T. rex*, but it can't be *T. rex* because you don't find *T. rex*." At the time only 11 other *T. rex* had ever been found.

T. rex actually lived closer in time to the first humans (about 60 million years apart) than it did to the first dinosaurs (about 160 million years apart). It first appeared toward the end of the Cretaceous period, probably about 67 million years ago. Its evolutionary history remains somewhat cloudy, but several respected scientists believe that it may have been closely related to a Mongolian meat eater, *Tarbosaurus bataar* (also called *Tyrannosaurus bataar*). It is possible that descendants of this dinosaur emigrated from what is now Asia to what is now North America; the continents were connected at that time.

The famous fossil collector Barnum Brown found the first three *T. rex* in the Wyoming and Montana badlands in the early 1900s while on expeditions for the New York-based American Museum of Natural History. Like the Larsons, Brown became interested in fossils at an early age, collecting specimens uncovered by the plow on his family's farm in Carbon Hill, Kansas. And like Hendrickson, he seemed to possess a sixth sense for finding bones, or at least a unique fifth one. "Brown is the most amazing collector I have ever known. He must be able to smell fossils," said Henry Fairfield Osborn.

It wasn't Brown's sense of smell but his vision that resulted in his first Montana find in 1902. Months earlier, William Hornaday, the director of the New York Zoological Society, had shown Brown a paperweight made from a fossil he had found while hunting in eastern Montana. Brown identified the fossil as dinosaur—part of a *Triceratops* horn. After looking at photographs of the area where Hornaday had been hunting, Brown sensed that the land, part of the Hell Creek Formation, might be ripe with dinosaurs.

He was right. In July 1902, Brown found dinosaur bones in a sandstone bluff on the same ranch where Hornaday had found his fossil. The skeleton, which contained about 10 percent of the creature's bones, was not fully excavated and shipped to New York until 1905. By that time Osborn had identified it as a new species of dinosaur, which he christened *Tyrannosaurus rex* ("tyrant lizard king").

In 1907, Brown discovered a second *T. rex* skeleton in eastern Montana. This specimen was even better than the first—with 45 percent of the bones, including an excellent skull—"a ten strike," in Brown's words. In between these finds, Osborn determined that bones Brown had found in 1900 in the Lance Creek Beds of Wyoming were also those of a *T. rex*.

The *T. rex* was the biggest carnivore discovered to date. It had massive legs, a short, thick neck, and a narrow chest. Its tail was relatively short, and its hips were relatively narrow. Its arms were surprisingly small, about the length of human arms—just 3 feet long. It had two fingers. Its huge skull was heavily reinforced, distinguishing it from other big carnivores. It had sharp, deadly teeth the size and shape of bananas. When those teeth fell out, new ones replaced them. The dinosaur itself may never have stopped growing.

As the biggest, baddest dinosaur, *T. rex* quickly captured the fancy of early twentieth-century America. The press dubbed it "The Prize Fighter of Antiquity" and hyperbolized that "the swift two-footed tyrant munched giant amphibians and elephant au naturel." In 1915, record crowds flocked to the American Museum when it mounted the first public exhibition of a *T. rex*, unveiling a fully restored, freestanding skeleton of Brown's second Montana find. As Philip Currie, curator of dinosaurs at Canada's Royal Tyrrell Museum of Paleontology, wrote in the preface to Sotheby's catalog for Sue, *Tyrannosaurus rex* became the "standard against which other dinosaurs are measured . . . the most famous dinosaur."

Most paleontologists consider a *T. rex*, which has approximately 300 bones, scientifically significant if it is at least 10 percent complete. Some, like Larson, determine this percentage based on the number of bones found. Others peg their figure to the percentage of the dinosaur's total surface area found, no matter how many bones are recovered. Proponents of each method agree that 59 years passed between Brown's last find and the discovery of the fourth *T. rex* in 1966 by Harley Garbani, a professional plumber and longtime amateur paleontologist collecting in eastern Montana for the Los Angeles County Museum of Natural History. Over the next 24 years, seven more *T. rex* would be found in Montana, South Dakota, and Alberta. The best of these specimens, the eleventh *T. rex*, was found in 1988 in the eastern Montana badlands by Kathy Wankel, a local rancher. Horner and a crew from the Museum of the Rockies excavated the skeleton in June 1990, just two months before Hendrickson made her discovery. Forty feet long and, according to Horner, almost 90 percent complete, this find dethroned— for the moment, anyway—Brown's American Museum *T. rex* as the finest ever.

When Hendrickson found what she thought might be the twelfth *T. rex*, she tried to contain her excitement. "I didn't jump up and down and scream," she says. "But I was thinking, Wow! What was so cool was that the vertebrae were mostly going into the hill, so it looked like the potential for more. Usually you find the last little bit of bone and there's nothing more [because it has eroded]. I knew this was part of one specimen, and that if the visible bones were all there was, it would have been important. But I knew there was more."

Hendrickson didn't disturb the bones in the cliff. She did, however, pick up a few pieces from the ground, each about 1½ to 2 inches across. "They were all hollow," she says. "I've picked up thousands of other pieces of bones before, and they were all solid." Theropods, the class of carnivorous dinosaurs that included *T. rex*, were hollow-boned, like birds. The excitement building, Hendrickson flew back to camp, where she knew Larson would have resumed excavating after the trip to town.

She found Larson on his knees digging up the *Triceratops* skull on Sharkey Williams's land. "Pete, I have to show you something," she said breathlessly.

"I had never seen *T. rex* vertebrae, but I knew that's exactly what I was looking at," says Larson. He is an unassuming man, trim with sandy hair and a thick mustache. Dressed in khaki slacks and a plaid shirt, he wears wire-rimmed glasses. He looks like an athletic academic as he sits in his small, book-filled office in the institute's basement remembering the discovery. Just down the hall, a technician is using a device called an airbrade to clean some dinosaur bones. The airbrade sprays powder with compressed air to blow off dust and rock from the bone. The basement sounds like a dentist's office.

How did Larson know Hendrickson had brought him *T. rex* vertebrae? "The size of the fragments, the curvature of the bone. The open spaces." He picks up a cervical vertebra from a *T. rex* found after Sue. "It's been waterworn before it was fossilized, and you can see where the surface of the bone has been weathered away," he explains. "You can see these open spaces inside, kind of a honeycomb texture. That's exactly the same texture you find in bird bones and theropods—where birds come from. This is all connected to the respiratory system through the openings right here. There are air sacs, just like birds. *I just knew what the fragments were.*"

Identifying fossils was not always so easy. In his delightful book *Dinosaur Hunters,* David A. E. Spalding, the former head curator of natural history at the Provincial Museum of Alberta, tells the story of one Robert Plot. Plot was a seventeenth-century British naturalist and the first curator of the Ashmolean Museum in Oxford. Like many scientists of the day, he was still influenced by the mysticism of the Middle Ages. He speculated that fossils might have been created by petrifying juices to adorn the inside of the earth, just as flowers had been made to adorn the surface.

In his 1677 opus *The History of Oxfordshire,* Plot provided an illustration of what appears to be a dinosaur bone. If so, it is the first published record of such a fossil. Of course, Plot didn't identify it as such; the word "dinosaur"—from the Greek *deinos,* meaning "terrible," and *sauros,* meaning "lizard"—was coined 165 years later in 1842 by the British paleontologist and anatomist Sir Richard Owen. Instead, Plot said that the bone, dug from a quarry and given to him by a fellow Englishman, was similar to the lowermost part of the thigh of a man "or some greater animal than either an Ox or Horse and if so it must have been the Bone of some Elephant, brought hither during the Government of the Romans in Britain."

After seeing a living elephant for the first time, Plot changed his mind. The bone must have belonged to some giant man or woman, he said. While the actual fossil has been lost, Spalding argues that the illustration is clearly that of a dinosaur bone, probably of the carnivorous *Megalosaurus.*

Fast forward almost 90 years to 1763 and the publication of Richard Brookes's *The Natural History of Waters, Earths, Stones, Fossils, and Minerals.* Brookes copied Plot's drawing and described it as *Scrotum humanum.* Writes Spalding: "It is not known whether this was a serious attempt at identification or a bizarre joke."

Brookes's description was taken seriously by J. B. Robinet, a French philosopher. In 1768, Robinet called the fossil a "stony scrotum." Such objects represented nature's attempts to form human organs in a quest to create the perfect human type, he hypothesized.

Eager to test his hypothesis that Hendrickson had found a *T. rex*, Larson hustled his colleague into the institute's truck. They sped to Maurice Williams's fence line, then ran the additional 2 miles to the cliff. "Here," Hendrickson said, pointing to the bones, "this is my going-away gift to you."

The normally mild-mannered Larson admits that he got excited. "Sue took me over to the spot, and there were literally thousands of little fragments of bone lying on the ground and some bigger pieces and you could see there were parts of vertebrae and we looked up about 7 feet up on the face of this cliff and there was this cross section of bones about 8 feet long coming out and I crawled up there and we could see three articulated vertebrae. *And I knew at that instant that it was all there.* Call it intuition or whatever. I just knew. I knew this was gonna be the best thing we'd ever found and probably ever would find."

After locating the cliff on a topographic map, Larson called his assistant, Marion Zenker, at the institute and asked her to verify that Maurice Williams owned the section of land on which the cliff was located. Williams had given permission to look on his property, but determining who owns what within the boundaries of a reservation is often problematic. Some land is owned outright by individuals. Some land is owned by the tribe. Some land is leased from the tribe by individuals. And some land owned by individuals is held in trust by the federal government. In some instances, one ranch "owned" by a Native American may include parcels of land that fall into all these categories.

Larson wanted to make sure he was dealing with the true owner of the site where the bones sat. Zenker called the land registrar for Ziebach County, where the ranch was located. The registrar told her that Maurice Williams owned the land and had a couple of oil leases on file.

Larson called Williams that night. "I told him we had found something that looked really good," Larson says. Williams gave him permission to excavate, but still forbade him to drive onto the property. "I told Maurice, 'It's big. We're going to have to drive on at least once to take it off,' and he said, 'Okay,'" Larson remembers.

That night dinosaurs visited Larson as he slept. Such visits weren't uncommon, especially when the paleontologist was on a dig. "I always dream about what I'm going to find," he says.

Larson wanted his brother Neal, who had gone home for the weekend, to see the site before the digging began.

"I was thinking I'd be coming back and we'd be closing up the quarry for the summer," Neal says. "Then I got a phone call from Peter on Saturday night."

"When you come back, I want you to bring lumber and plaster of Paris, and, oh, bring the trailer, too," Peter Larson said as nonchalantly as possible.

"What did you find, Pete, a skeleton?"

"You'll see."

Neal was hooked now. "What kind of skeleton? Is it a *Triceratops?*"

"You'll see when you get here."

"Is it a duck-bill?"

"You'll see." Older brothers didn't torment younger brothers in the days dinosaurs roamed the earth—but only because there weren't any humans around then.

Neal Larson was eager to learn what kind of skeleton had been found, but he wasn't eager enough to leave for the quarry the next day. "It was Sunday, and I hadn't been to church for a few weeks," he explains.

Neal, a little rounder and a little less studious looking than Peter, arrived in Faith on Monday, August 14, with the trailer and all the other supplies on Peter's shopping list. By that time, Peter had christened the dinosaur Sue, after Hendrickson, who still has mixed emotions about the appellation. "I'm deeply honored," she says. "It's just that I've never liked my name—Sue, Susan, whatever. I just don't care for it." She adds, "Of course, at the point Pete named it after me, it was just three articulated vertebrae. We didn't know how great she'd be."

Before Neal's arrival, Larson had also videotaped and shot still photos of the find and had taken the other members of the team to the cliff, including Terry Wentz, the institute's chief fossil preparator, and the two boys, Matthew and Jason Larson. During their weeks at the quarry, Matthew had found more than a dozen teeth apparently shed by *T. rex*. Hendrickson hadn't found any, a fact that Matthew never let her forget. Responding to his good-natured teasing, Hendrickson had laughed: "Matt, I just want to find the whole thing." Now they would see just how whole the skeleton was.

Peter Larson walked Neal to the cliff, and said, "Gee, can you tell me what it is?" He was still playing the big brother.

"Well, it's big."

"Yeah."

"Is it *T. rex*?" asked Neal, who had never seen such bones before.

Just like the dinosaur itself, the site of a find is of great scientific importance. It may contain fossilized plants or bones from other ancient creatures that provide a context for the dinosaur's life. Good paleontologists and fossil hunters don't just excavate a find, they harvest the rock around it as well.

Example: In 1996 John Flynn, head of the Field Museum's geology department, returned from a dinosaur dig in Madagascar with, among other things, a 50-pound bag of dirt collected at the site. Over the next three years, museum volunteers sifted the dirt through screens and then used microscopes to look for fossils. In 1999, the museum announced that the dirt had yielded a 165-million-year-old fragment of the jaw of a mouse-sized mammal. The fragment, only half the length of a grain of rice and containing three teeth invisible to the unaided eye, proved for the first time that mammals were alive and sharing the world's southern continents with dinosaurs far earlier than previously believed.

Larson's crew began by picking up all the scraps from the ground and putting them in plastic bags, which they carefully labeled. They then bagged much of the surrounding dirt for future screening. Next, they stabilized the bones sticking out from the cliff with a hardener and covered them with burlap and plaster of Paris—a technique first described by the English geologist Sir Henry Thomas de la Beche in 1836 and perfected in the late nineteenth century by those hunting bones in the American West.

This was easy compared to the next task: removing 29 feet of over-burden. For five days the Larson brothers, Hendrickson, and Wentz worked with pick and shovel to clear the sandstone and hard soil above the skeleton. "These were the hottest days of the summer," says Hendrickson. "The temperature was 115 plus. You're trying to find shade, but there is none. And we don't stop at noon."

They dubbed the site "Tyrable Mountain."

Once down to the level of the skeleton, the team would use knives, brushes, and smaller tools to remove the bones. But before this removal could begin, most of the bones required special treatment. Those in danger of cracking were glued on the spot with commercial Superglue. Those in danger of crumbling were squirted with a liquid solution that hardened them.

Documentation of a dig is critical. Scientists studying a specimen in the lab want to know where each and every part was found in the field. Larson's team used a mapping technique learned from Bakker, marking the location of each bone by tracing it full scale on butcher paper. They also took still photos and videotape.

Maurice Williams visited three or four times during the excavation. A tall, sturdy man, he is seen on the videotape in sunglasses, wearing a T-shirt and baseball cap. On one occasion, the crew took a break and helped him dehorn some cattle. On another occasion, the institute camcorder captured the following conversation between the rancher and Peter Larson:

Williams:	You are going to mount her in Hill City.
Larson:	Yeah.
Williams:	Good. *(Pause)* And under that you'll write, "Stolen from Maurice Williams."

Williams, Larson, and the others at the site all laughed at this last line.

The Larsons quickly determined that Sue was large for a *T. rex*. Her 54-inch femur suggested that she would have stood 13 feet tall at the hip and 41 feet long, a foot taller and 2 to 3 feet longer than Barnum Brown's famous *T. rex* at the American Museum of Natural History. "At that size, running at 25 to 40 miles an hour, this was one big, terrible dinosaur," says Neal Larson.

Brown's lengthy excavation of the first Montana *T. rex* in the pre-truck era presented numerous logistical problems. Encased in plaster, the dinosaur's pelvis alone weighed 4000 pounds, far too much for a conventional horse-drawn wagon. What to do? Brown built a wooden sledge and then hitched it to a team of four horses for the 125-mile journey to the closest railroad station.

Excavating Sue was not nearly so problematic or time-consuming. Because she was so complete and her bones were confined to such a small area (about 25 feet by 30 feet), the crew was able to dig her out in only 17 days. Still, they suffered several anxious moments—actually, several anxious days.

Larson had immediately sensed that Sue "was all there." But was she? A week after removing the overburden, her skull—the most important part of her body from both scientific and display standpoints—still hadn't surfaced. Then, on August 27, Larson hit something with his pick. "It's the S word," he excitedly told his crew. ("I didn't say 'skull' because I didn't want to jinx it," he later explained.) Larson's fellow diggers were skeptical.

The paleontologist took pick to rock and again hit something hard. "I tell you it's the S word.'" But his crew remained skeptical until he found the curve of the cheek some hours later.

The good news was that they'd found Sue's skull. The bad news was that the nose was buried under Sue's hips, all 1500 pounds of them. "It seemed that it had stuck its head between its legs and kissed itself good-bye," Neal Larson later said with a laugh.

Putting an end to this 67-million-year good-bye kiss on site would have meant saying good-bye to the skull. It was sure to be severely damaged by any attempt to extricate it from the pelvis. The crew decided to remove these bones as a unit in a plaster-jacketed block. Then, back at the institute, the Larsons and Wentz could figure out the safest way to separate them.

Peter Larson says that he told Williams he would pay him for the fossil "early on" after the discovery. Finding the skull triggered "negotiations." Says Larson: "After we could see what was there, I told Maurice, 'This is a really good specimen. I'll give you $5000 for it,' and he said, 'Fine.' I wrote on the check what it was for."

The check was marked "For theropod skeleton Sue/8-14-90-MW." Williams deposited the check the day after he received it.

Excavated bones were wrapped in foil and then jacketed in plaster. By August 31, the crew, which had grown to six, had removed everything except for the big block containing the skull and pelvis, sacrum, several dorsal vertebrae, ribs, the right leg, some foot bones, and, they hoped, the right forelimb. With continuous undercutting and plastering, they had this block ready to be removed by 5:00 PM on September 1. Because a large rainstorm was heading their way, they decided not to wait overnight to pull the 9-foot-by-7-foot 9000-pound block from the ground and onto the trailer. Three hours later the trailer and three trucks carrying another 5 tons headed back to Hill City. There, the institute would begin the task of readying Sue for study and display.

After spending the night in Hill City, Peter Larson and Sue Hendrickson climbed into their respective trucks and drove in tandem to Bozeman where they showed the pictures of their find to the Museum of the Rockies' Horner and discussed the fossil's future—a future that at the time looked quite rosy.

The couple's future was much bleaker. Although they still cared deeply for one another and had just collaborated on what appeared to be one of the greatest dinosaur finds in history, they remained committed to ending their romance. Twenty-four hours after reaching Bozeman, Hendrickson bid Larson good-bye and left to visit her family in Seattle. She and Larson were sure they'd see each other again, but never in their wildest dreams did they imagine the bizarre chain of events that would soon bring them face to face and cause them both such pain.

Although she felt a deep kinship with her namesake, Hendrickson sought no remuneration for her efforts. The thrill of discovering the best *T. rex* ever was enough. And besides, Larson had said that he wanted to build a new museum in Hill City and make Sue its star attraction. No one had ever gotten rich finding a fossil, Hendrickson told herself, and no one was going to get rich because of this dinosaur.

2

NEVER, EVER FOR SALE

"Six hundred thousand dollars," Redden said.

The auction floor was abloom in blue paddles.

"Sevenhundredthousandeighthundredtousandninehundred-thousand." Redden had all he could do to harvest the bids.

Sue Hendrickson sighed. Grounded in reality, she sensed that Peter Larson's angel might not have enough money to win back the *T. rex*.

A fading color photograph, circa 1960, hangs on a wall at the Black Hills Institute. The picture is set in a farmyard, where four young children—three boys and a girl—stand by a table full of bones in front of a small wooden building. On a nearby post that is adorned by a bleached cow skull, a cardboard sign lettered in a child's hand identifies the building. "MUSEUM," it boldly proclaims.

Three weeks after bringing Sue home to Hill City, two of the boys in the photograph, now grown men, announced in the *Rapid City Journal* that they had found what they believed to be the largest, most complete *Tyrannosaurus rex*. Peter and Neal Larson added that they intended to make the dinosaur the star attraction in a new museum—a museum they had dreamed of building for more than 30 years. "The largest meat eater ever to walk the face of our planet would surely draw visitors to our door," Peter said. The *T. rex* will "never, ever" be for sale, Neal told the newspaper. Few newspapers or magazines outside South

Dakota carried the story of the find or the plans for the museum; Sue may have been one of the biggest fossils in the world, but she wasn't famous outside greater Rapid City . . . yet.

The brothers explained that Sue appeared to have a complete pelvis; nearly complete torso and tail; the radius, ulna, and hand bones of the small arms; the first complete *T. rex* shoulder girdle ever found; possible stomach contents; and a spectacular skull that included a lower jaw and a set of "dangerously serrated, dagger-like teeth." They speculated that she might have weighed up to 6 tons. Sue showed evidence of having indeed been the prizefighter of antiquity, the Larsons told the *Journal.* Her head had been injured. One leg appeared to have been broken and then healed. Vertebrae in her tail were fused. These battered bones and her large size indicated that she had "led a long and perilous life," Peter reported. He would later say that he was surprised that they hadn't found a wheelchair buried beside her.

Larson found these injuries fascinating. "I'm interested in putting 'flesh' onto the skeleton," he says. "A fossil's pathology can give us a snapshot of its day-to-day life." He pictured Sue, her huge leg broken, resting in tall ferns, unable to move very far. This led to the question How did she get enough food to survive until the leg healed? Larson pictured a caring mate or clan members bringing her food.

The injuries to Sue's head suggested another series of snapshots. In his self-published book *The Rex Files,* Larson visualized a battle between Sue (the Matriarch) and another female *Tyrannosaurus rex* for control of the clan or territory:

> The younger theropod was not as strong or as large as her adversary, but she was more agile. The Matriarch was well seasoned for combat. She had repeatedly defended her position in the group. . . . Her neck pulsed with pain as she wheeled on her attacker. This injury, a bite received more than ten years ago from a pretender to the throne, had never healed, but continued to ooze putrescence.
>
> The quickness of the challenger, nearly 50 years the junior of the two combatants, saved her from the savage attack as she moved effortlessly to the right. With incredible speed, she caught her elder off balance and attacked her from behind. The

younger Tyrannosaur sunk her serrated, dagger-like teeth into the left side of her opponent. Bone shattered behind the eye of the Matriarch. A tremendous scream filled the air. . . . The mighty ruler fell face down in the stream as blood gushed from the fatal wound and mixed with the muddy waters. . . .

Slowly, ever so slowly, the stream covered this once mighty body while organisms stripped the bones of its flesh. Sand, leaves, and other bones accumulated around and on top of the skeleton. Within weeks the burial was complete.

Sue was always a female in Larson's snapshots and conversations. "We'd given the *T. rex* that name and always referred to it as 'she' or 'her,'" he says. "But even at the site, people were asking me, 'What if she's a male?' I'd say that we could always point to that Johnny Cash song, 'A Boy Named Sue.'"

Actually, Larson had little reason to believe that this *T. rex* was a girl named Sue. At the time, most scientists, with the notable exception of Ken Carpenter of the Denver Museum of Natural History, had speculated that the smaller, "gracile" *T. rex* were female, while the larger, more "robust" *T. rex* were male. And Sue was the largest, most robust *T. rex* ever found. Further study of "her" bones—and the bones of the Museum of the Rockies' specimen, which appeared gracile—might finally provide the answer to determining gender.

The bones might also answer another question that had long been a source of spirited debate in the scientific community: Was the *T. rex* primarily a predator or a scavenger? Years earlier, those in the scavenger camp had argued that the *T. rex* was a fat, sluggish, cold-blooded creature, too slow to catch its prey and too big to sneak up on it. By the time the Horner and Larson parties had excavated their dinosaurs in 1990, however, most agreed that the long-legged *T. rex* possessed the speed to run down its fellow dinosaurs. But what about its almost comically short arms? Were these stubby appendages, only 3 feet long, strong enough to grab a meal on the run?

Perhaps, said Carpenter and Matt Smith, a graduate student who also worked at the Denver Museum. The January 1991 issue of *Discover* magazine reported that the pair had measured a scar on the left forelimb of the Horner *T. rex* where the tendon was attached to the biceps.

"Given the size of the bone and the angle of the tendon attachment, and estimating the mass of the muscle, they calculated that each of the animal's arms could have withstood a weight of 1200 pounds." This was more than enough to bring down prey. "I don't know what a scavenger would be doing with arms like that," Smith told the magazine.

Maybe making love. "I think it's too soon to say that the arms were used to grasp prey," Horner told *Discover*. "These arms are so short, *T. rex* would have had to put its chest down on the prey to stabilize it. I'm not sure other dinosaurs would have been so slow and stupid as to let that happen." He would later write that earlier theories that the arms were used to hold or tickle the mate during foreplay seemed more plausible.

Larson didn't know exactly what the arms were used for, but that didn't matter in answering the predator or scavenger question, he asserted. One look at Sue's powerful jaws—Bakker compared them to railroad spikes—persuaded Larson that she didn't need arms to hold her prey. "She was an opportunistic feeder," he hypothesized. If she came upon an animal that was dead or injured, she wouldn't hesitate to eat it. But if she were hungry and there was nothing to scavenge, she'd seek out a smaller dinosaur, run it down, and grab it with her choppers, just as a wolf does. "Her mouth was her arsenal. She had a skull full of bullets," Larson said. Study of her skull would, he hoped, shed additional light on this subject. Analysis of the olfactory lobe would, for example, determine if the *T. rex* had the sense of smell necessary for identifying, discriminating, and tracking prey.

But what difference does it make if Sue had a caring mate or if "she" was male or female, predator or scavenger? What can the study of fossils in general and dinosaurs in particular tell us about our world?

For his superb book, *The Riddle of the Dinosaurs,* Pulitzer Prize–winning science writer John Noble Wilford unearthed a telling 300-year-old quotation from the seventeenth-century English experimental scientist Robert Hooke. Hooke was a true post-Renaissance man. He constructed the first Gregorian telescope, discovered plant cells, was the first to use spiral springs to regulate watches, stated the currently accepted theory of elasticity (Hooke's Law), and even recognized some aspects of the law of gravitation before Sir Isaac Newton. He also studied fossils and concluded:

I do humbly conceive ('tho some possibly may think there is too much notice taken of such a trivial thing as a rotten shell, yet) that men do generally too much slight and pass over without regard these regards of antiquity which Nature have left as monuments and hieroglyphick [sic] characters of preceding transactions in the like duration or transactions of the body of the Earth, which are infinitely more evident and certain tokens than anything of antiquity that can be fetched out of coins or medals, or any other way yet known, since the best of those ways may be counterfeited or made by art and design. . . . [Fossil shells] are not to be counterfeited by all the craft in the world, nor can they be doubted to be, what they appear, by anyone that will impartially examine the true appearances of them: And tho' it must be granted that it is very difficult to read them, and to raise a chronology out of them, and to state the intervalls [sic] of the times, wherein such or such catastrophes and muta-tions have happened; yet 'tis not impossible.

Three centuries after Hooke's observations, the Polish paleontologist Zofia Kielan-Jaworowska explained the rationale for studying dinosaurs in her 1969 book, *Hunting for Dinosaurs:* "The study of animals that lived on earth millions of years ago is not merely a study of their anatomy, but first and foremost a study of the course of evolution on earth and the laws that govern it."

The first comprehensive study of the dinosaurs that lived on earth mil-lions of years ago was presented by Sir Richard Owen. In 1842, Owen delivered his "Report on British Fossil Reptiles" to the eleventh meeting of the British Association for the Advancement of Science. In the report, he not only introduced the term "dinosaur" to the English language, he introduced the notion of the "terrible lizard" itself, distinguishing it from all other reptiles past and present.

Owen's analysis was triggered by two presentations in the 1820s about new fossil finds. In 1824, William Buckland, an English minis-ter/geologist to whom "rock of ages" apparently had a double meaning, published a paper after studying the bones of a creature christened *Megalosaurus,* or "great lizard." (By this time scientists had begun to use

Greek or Latin words to describe their finds.) Buckland's contemporaries speculated that this carnivore found in a quarry at Stonesfield some ten years earlier was 40 feet long and 8 feet tall.

Among those in attendance when Buckland presented his paper on *Megalosaurus* to the Geological Society was Dr. Gideon Algernon Mantell, an English surgeon with a love for collecting fossils. Within months, Mantell would write his own paper on the discovery of a new herbivorous reptile. In later years, the doctor gave credit for the find to his wife, who shared his paleontological passion. Mary Ann Mantell, the story goes, accompanied her husband on a house call to Sussex. As the doctor attended to his patient, Mrs. Mantell went for a walk, whereupon she found some fossil teeth. While there seems to be some question about the specific facts of this story, there is little doubt that in 1820 or 1821 Mantell or his wife had found the teeth of what would prove to be a dinosaur.

How is a new extinct reptile identified? In this case, through dental records. But it was not an easy job. Mantell took his fossils to several people before receiving satisfaction. Members of the Geological Society were less than excited; they speculated that the teeth belonged to a large fish or mammal. Next, the great French paleontologist, Georges Cuvier, father of the theory of extinction, dismissed them as rhinoceros incisors.

Undaunted, Mantell went to the Hunterian Museum at the Royal College of Surgeons in London. After failing to find any similar reptilian teeth in the museum's collection, Mantell happened to meet a young assistant curator who was studying iguanas. Samuel Stuchbury observed that Mantell's fossil teeth closely resembled the teeth of a living iguana found in Central America—only they were considerably larger. When Stuchbury showed him an iguana skeleton, Mantell agreed.

The doctor finally had an answer. The teeth were neither fish nor fowl nor rhinoceros, but rather belonged to a long-gone, giant plant eater. He named this reptile *Iguanodon* ("iguana tooth"). Based on the fact that the living iguanas of the day reached 5 feet in length, Mantell estimated that the *Iguanodon* might have been 200 feet long. He soon revised the length to 60 feet, which was still three times too great.

By the time Owen presented his famous paper 17 years later, Mantell's herbivorous *Iguanodon* and Buckland's carnivorous

Megalosaurus had company. Several more names had been added to the pantheon of terrible lizards. The latest seven nondwarfs were *Cetiosaurus, Cladeidon, Macrodontophion, Palaeosaurus, Plateosaurus, Thecodontosaurus,* and *Hylaeosaurus* (an armored dinosaur also first described by Mantell).

These lizards shared so many traits that Owen proposed "establishing a distinct tribe or suborder of Saurian Reptiles . . . *Dinosauria.*" The paleontologist noted several ways in which these extinct animals had differed from the other saurians. They lived on land rather than in the water. Unlike other reptiles, they all had five vertebrae fused to the pelvic girdle. And they were bigger and more massive, their bodies somewhat reminiscent of the elephant and its fellow pachydermal mammals.

Encouraged by Britain's Prince Albert, Owen eventually brought this vision of the dinosaur and its prehistoric surroundings to life for the masses. When the Crystal Palace moved to Sydenham in South London in 1853, the prince commissioned Benjamin Waterhouse Hawkins, a painter/sculptor, to, in his words, "summon from the abyss of time and from the depths of the earth those vast forms and gigantic beasts which the Almighty Creator designed with fitness to inhabit and precede us in possession of this part of earth called Great Britain." With technical assistance from Owen, Hawkins created life-size versions of *Megalosaurus, Iguanodon,* and their fellow inhabitants of ancient Great Britain.

Later that year, Owen and 20 other leading natural scientists gathered at the Crystal Palace for a New Year's Eve party. The theme was dinosaurs, and the dinner itself was held inside a huge, life-size model of an *Iguanodon*—perhaps the first theme restaurant. The guests enthusiastically sang a playful homage to the dinosaur written by Professor Edward Forbes of the Museum of Practical Geology:

> A thousand ages underground
> His skeleton had lain,
> But now his body's big and round,
> And he's himself again!
>
> His bones, like Adam's wrapped in clay,
> His ribs of iron stout,

Where is the brute alive today
That dares to turn him out?

Beneath his hide he's got inside
The souls of living men;
Who dare our saurian now deride
With life in him again?

(Chorus)
The jolly old beast
Is not deceased.
There's life in him again.

Forbes instructed the singers to conclude the chorus with a "roar."

Owen was much better at describing dinosaurs than at designing them. Because he was working with incomplete specimens, the man rightfully deemed the "father of paleontology" wrongfully conceived his terrible lizards as more like mammals than reptiles. As a result Hawkins's beasts were not entirely anatomically correct. His *Iguanodon*, for example, had a horn on its snout. In reality, that horn was a dagger-like spike, and it should have been located on the hand. Some years later, the American paleontologist Othniel C. Marsh poked fun at the Owen/Hawkins models. "There is nothing like unto them in the heavens, or on earth, or in the waters under the earth," he said. "The dinosaurs seem to have suffered much from their enemies and their friends."

The Almighty Creator's gigantic beasts had possessed more than Great Britain. Accordingly, in 1868 the commissioners of New York's Central Park invited Hawkins to create life-size models of the creatures that had inhabited ancient North America. The commissioners planned a Paleozoic Museum for the park as well. Unfortunately, when Boss Tweed grabbed control of New York City, the project was killed. All work done to date was buried beneath the park, to be unearthed perhaps by future fossil hunters.

Hawkins's star attraction was to have been the *Hadrosaurus foulkei* ("Foulke's big reptile"), a Cretaceous-period duck-billed dinosaur. William Foulke, a member of the Philadelphia-based Academy of

Natural Sciences, and Joseph Leidy, an anatomist and the academy's director, had unearthed *Hadrosaurus* in 1858 on a farm in Haddonfield, New Jersey. Earlier that year, Leidy had been the first in America to identify fossils as dinosaurian. After examining teeth found by a U.S. Army mapping expedition in the badlands of what is now Montana, Leidy declared them to belong to two different dinosaurs. One was a duck-billed herbivore; he named it *Trachodon* ("rough tooth"). The other was a carnivore; he named it *Deinodon horridus* ("most horrible of the terror teeth"). Such efforts led many to regard Leidy as the "father of vertebrate paleontology in North America." (Science seems to have fathered an entire genus or species: "father of . . . .").

When fully excavated, the Haddonfield bones turned out to be the best dinosaur skeleton found up to that point in history. The specimen was 30 feet long. Its fore and hind limbs were intact, as was its pelvis. It included 28 vertebrae, 9 teeth, and fragments of the jaw.

The teeth of "Foulke's big reptile" were similar to those of Mantell's *Iguanodon*, which Owen had portrayed as rhinoceros-like. Blessed with a more complete skeleton than Owen had to work with, Leidy visualized a somewhat different animal. He told the Academy of Natural Sciences that the thigh bone was so much longer than the humerus bone of the upper arm that if they had not been found together, he would have thought they came from different dinosaurs. What did he make of this? "The great disproportion of size between the fore and back parts . . . leads me to suspect that this great extinct herbivorous lizard may have been in the habit of browsing, sustaining itself kangaroo-like in an erect position on its back extremities and tail."

Evidence of a bipedal dinosaur seemed to answer a question that had puzzled lay people and scientists alike for almost 60 years: Who left the numerous giant, three-toed tracks found throughout the Connecticut Valley? The footprints seemed too big to have come from birds, but until the discovery of *Hadrosaurus foulkei* there was no other candidate.

Creationism created some of the confusion. Witness the case of young Pliny Moody. While plowing his father's farmland near South Hadley, Massachusetts, in 1802, Moody uncovered a slab of sandstone that bore a three-toed footprint similar to that of a turkey or raven, but much bigger. The good religious people of the area concluded that the print belonged to the raven that Noah had sent forth from the ark to find terra firma.

Following in Moody's footsteps was Edward Hitchcock, a nineteenth-century Congregationalist minister who taught chemistry and natural history at Amherst College and eventually served as that institution's president. Hitchcock became the leading expert on fossil footprints of the eastern United States, collecting them, studying them, and classifying them with descriptive Latin names. Many of these tracks were found in the sandstone quarried to build the brownstone townhouses of early nineteenth-century New York. None were accompanied by skeletons. As a result, Hitchcock could only speculate about the creatures that had left them.

In 1848, after 13 years of study, Hitchcock summarized his findings in a paper for the American Academy of Arts and Sciences, "An Attempt to Discriminate and Describe the Animals that Made the Fossil Footprints of the United States." The Connecticut Valley, he wrote, had long ago been home to huge three-toed birds that were several times larger than ostriches.

While turn-of-the-nineteenth-century creationists followed the Connecticut Valley's three-toed fossil tracks back to Noah's ark, mid-century evolutionists followed those same tracks to Darwin's *Beagle* after Leidy's postulation of a bipedal dinosaur. Leading the march was Thomas Henry Huxley, a brilliant scientist who had eschewed a career in medicine for one in paleontology and natural history.

Dubbed Darwin's "bulldog," Huxley took to the *Hadrosaurus* like a dog takes to a bone. "The important truth which these tracks reveal is that, at the commencement of the Mesozoic epoch [from which the sandstone dated], bipedal animals existed which had the feet of birds, and walked in the same erect or semierect fashion," he argued.

Why was this truth so important to the evolutionists? The survival of Darwin's theory, advanced in his 1859 landmark, *On the Origin of Species by Means of Natural Selection, or the Preservation of Favored Races in the Struggle for Life*, was in no small measure dependent on the ability to find missing links. For example, from what animal had the bird evolved? Darwin had hypothesized that reptiles were the link. The discovery that dinosaurs had left the bird-like tracks in America seemed to support this hypothesis.

Another fossil added even more fuel to the evolutionists' fire. In late 1860 or early 1861, stonecutters working in the limestone beds of Bavaria found the remains of a strange creature; it looked like a reptile, but it had a feather. If it was a bird, it was the earliest one on record.

Mid-nineteenth-century scientists weren't able to accurately date fossils, but they could determine the order in which they had appeared on earth based on the strata in which they were found. No bird had ever been discovered in limestone beds (which we now know formed about 150 to 160 million years earlier, during the Jurassic period). As a result, some scientists thought the specimen a hoax. But by August 1861, Hermann von Meyer, an eminent paleontologist from Frankfurt, had confirmed that the feather was genuine, though "not necessarily derived from a bird."

One month later von Meyer viewed a headless skeleton found along with the impression of feathers in the same limestone. "The fossil, indeed, looked like one of Darwin's missing links," writes Wilford. The skeleton had a long bony tail, three clawed fingers, and lizard-like ribs and vertebrae. Von Meyer gave the specimen a neutral name—neither bird nor reptile: *Archaeopteryx lithographica* ("ancient feather from the lithographic limestone").

Whether or not the *Archaeopteryx* linked birds to reptiles (most scientists today agree that it does), it did link science to commerce. The fossil's owner, Karl Häberlein, quickly announced that it was for sale. His asking price? The substantial sum of 700 pounds.

Peter Larson was not about to part with his find of a lifetime. "The Black Hills Institute's *Tyrannosaurus rex* will always be available for study and research since it is not for sale," Larson announced at the Fifty-first Annual Meeting of the Society of Vertebrate Paleontologists (SVP) in San Diego, California, in October of 1991. His speech and slide show, the first public presentation on Sue since her discovery, lasted 12 minutes—the maximum time the SVP granted presenters at that year's meeting. With only 12 minutes to summarize the first 12 months of research on only the twelfth (and best) *T. rex* ever found, Larson limited himself to only his most "startling discoveries": that this was the first *T. rex* skeleton with a nearly complete tail; that duck-billed dinosaur bones found with Sue showed etching and were coated with ironstone, suggesting that "this material is *actual stomach contents*"; that the skeleton showed evidence of a number of healed injuries; that the remains of three other smaller *T. rex* were found with Sue; and that Sue and the *T. rex* found by Kathy Wankel in Montana possessed very different body types. Larson said he was confident that over the next 18 months, "the completeness of Sue and the presence of other Tyrannosaur material will yield much new information on this largest of all land carnivores."

Larson then did something somewhat extraordinary: he invited any paleontologists interested in studying Sue to come to Hill City. Not all scientists and scientific institutions are so generous. Understandably, those expending a great deal of time, effort, ingenuity, and, money to find or secure a special specimen feel the need to get the best return on their investment. Often that return is realized through the publication of the first definitive groundbreaking scientific paper or papers on some or all aspects of the fossil. Thus, until the fossil has been studied and the paper written, these scientists or institutions will often deny meaningful access to outsiders. In this sense, science is similar to industry, which also jealously guards its new products until they are ready for the public.

Peter Larson does not believe in trade secrets. "We've always had a policy of being open," he says, explaining his decision to invite those who might be considered his rivals to come and study Sue and jointly publish a monograph to be called "The World of *Tyrannosaurus rex*." "Besides, there was more there to study than I could do in a lifetime." Within a few months, 34 scientists from around the world would sign on.

When Larson issued his invitation, several paleontologists had already traveled to Hill City. Visits by Currie of the Royal Tyrrell Museum of Paleontology and Carpenter of the Denver Museum of Natural History had quickly yielded dividends. Each had identified additional bones of the three smaller *T. rex* found with Sue. Another picture was forming in Larson's mind. "Does the presence of four Tyrannosaurs of such varied sizes in the same deposit mean that we found Mom, Dad, Junior, and Baby?" he asked at the SVP convention.

If all Larson had to offer was an exhibition of such impressionistic snapshots, he might rightfully be accused of practicing soft science. But for every photo in his mind's eye, there is a scientific paper in his name. Not only did he visualize Sue as a female, for example, he cowrote, with German scientist Eberhard Frey, an article for the *Journal of Vertebrate Paleontology* with the decidedly unsexy title "Sexual Dimorphism in the

Abundant Upper Cretaceous Theropod *Tyrannosaurus rex*." Translation: Let's look at the skeletons of a single species of dinosaur and see if they possess differences that are the result of being male or female.

How did Larson the hard scientist go about this research? He read everything remotely related to the subject that he could get his hands on. He journeyed across North America to see all available *T. rex* specimens. And he traveled to Karlsruhe, Germany, to consult with Frey, an expert on dinosaurs and on a reptile often used as a model for dinosaur behavior, the crocodile.

The living crocodile, like the extinct theropods, can be divided into two body types: heavy/robust and light/gracile. The heavy/robust crocodiles are male. These reptiles don't have a visible sex organ; it retracts into the body when not in use. The males are distinguishable from females, not only because they are larger but because they have an extra chevron (chevrons are the bony spines attached to the base of the caudal, or tail, vertebrae).

Just as the robust (male) crocodile possessed an extra chevron, so, too, did the gracile theropod. Therefore, Larson concluded: "In the theropod, the gracile body was male, the more robust, female. In all likelihood, the female was larger than the male."

In addition to studying crocodiles, Larson had also studied birds, thought to be the theropods' closest living relative. It appears that in all species of predatory birds, the female is larger than the male. These species appear to have something else in common: monogamy.

The paleontologist then returned to his snapshot. "By drawing a parallel with the predatory bird, I most unexpectedly came to believe that *T. rex* was probably monogamous and maintained family groups." He added that this possibility was supported by additional data: the fragmentary remains of a gracile male adult, a juvenile, and an infant with the robust female—Sue. Mom, Dad, Junior, and Baby. Dad, Junior, and Baby may have fallen victim to the same *T. rex* that killed Sue, Larson speculated. Perhaps Sue's clan had invaded the territory of another clan and a giant battle had ensued.

Bakker, for one, is impressed. "Pete is a wonderful scientist," he says. "He teased Sue's personality out of those bones."

Larson hoped to tease even more from Sue's skull. He was particularly interested in seeing where the nerves that ran through her brain and skull exited; the trail to the optic nerve might, for example, reveal clues to her visual acuity. Study of the inner ear and olfactory lobe and olfactory turbanals (tiny coils of bone in the nose) would offer more information about her sensory capabilities. The size of her braincase would reveal how big her brain may have been. Examination of the respiratory turbanals and holes in Sue's skull might also solve a question that had long perplexed scientists: Were dinosaurs cold-blooded or warm-blooded?

In Barnum Brown's day it would have been impossible to get this information without breaking apart the individual skull bones. Such bones are very hard to reassemble. "Breaking them apart is not a risk you want to take with the few *T. rex* skulls in existence," says Horner.

Fortunately, thanks to modern technology, neither Horner nor Larson had to take that risk. A simple CAT (computerized axial tomography) scan can photograph microscopically thin sections of bone and then reassemble the images in a three-dimensional view without damaging the skull. Well, maybe not simple. A 5-foot long skull won't fit in a conventional hospital CAT scanner.

Large scanners are almost as rare as *T. rex*, but considerably easier to find. Horner arranged for his *T. rex* skull to be scanned at the General Electric laboratories in Cincinnati, Ohio. There, GE x-rayed large products for defects. Larson shot for the moon. Bakker and Andrew Leitch, a dinosaur specialist from Toronto, spent months negotiating with the National Aeronautics and Space Administration for access to the same kind of scanner used by NASA to look for flaws in equipment like the space shuttle engines.

When these negotiations began, Sue's nose still lay under her massive pelvis in the preparation area at the institute. Chief preparator Wentz had begun the process of restoring and preserving Sue's bones, but he was saving the biggest for last. He realized that safely extricating the skull would be like playing a high-stakes game of pickup sticks.

There is no prep school for this kind of work. Wentz, a thin, balding man now in his mid-forties, learned his craft on the job. Before entering the fossil field, he worked in his family's cab business in southern Wisconsin. He has also worked in a flower shop and sold advertising. In 1984, a friend told Wentz about a notice posted at the University of Wisconsin inviting people to dig dinosaurs. A few weeks later Wentz found himself beside Peter Larson at the Ruth Mason Quarry. He ended up staying seven weeks. "I liked working outdoors, liked the fact that you never know what you're going to find next," he explains. "And I seemed to have a knack for it." The duck-bill bones Wentz and his fellow volunteers found were eventually reconstructed and displayed at the university in Madison.

In 1987, Wentz accepted Larson's offer to work full time at the institute and moved to Hill City. Soon he was preparing fossils in the lab as well as hunting them at the quarry. "Working in the field is great training for being a preparator," he says. "A bone is most fragile when you first expose it, so if you can get it out of the ground safely, you can probably do the work in the lab."

If you have the temperament. Here's how Wentz prepared Sue for her debut: He began by cutting open and removing the plaster field jackets from her bones, just as a doctor might remove a cast from a patient's limb. Then he peeled off the aluminum foil that had protected the bone from the plaster. Now it was time to remove the rock from the bone. He started with an air scribe—a minijackhammer with a point the size of a pen—then moved to hand tools like dental probes, paintbrushes, and toothbrushes. Finally, he pulled out a Micro-airbrade—a miniature sandblaster—which blasted the remaining rock off the bone with baking soda instead of sand. If the bones were cracked, Wentz used Superglue to repair them. He filled any gaps in the bone with a brown, clay-like epoxy putty that would soon harden.

Wentz acknowledges that many people might find cleaning a fossil as tedious as cleaning a house. To him, however, the wonders never cease. "No one looks closer at a fossil than the preparator does," he says. "I see it millimeter by millimeter—the valleys, the ridges. Each bone is like a whole other world."

Each bone is like a piece of a jigsaw puzzle, too. All the bone fragments have to be identified and pieced (glued) together. Larson, who is particularly skilled at looking at a tiny shard and knowing exactly where it belongs, sings Wentz's praise: "Terry is one of the finest preparators in the world today."

As work on Sue progressed, the Larsons and institute partner Bob Farrar worked on plans for the new museum. On March 13, 1992, they announced the creation of a nonprofit corporation to build what they would call the Black Hills Museum of Natural History. A fund-raising campaign had already begun. The Western Dakota Gem and Mineral Society had donated $1000, while students at Badger Clark Elementary School contributed $624.37.

Sue, of course, would be the centerpiece of the proposed 200,000-square-foot space. The Larsons estimated that it would take another 18 months to finish cleaning her bones and another year after that to make casts and assemble a life-size replica of her skeleton. In the meantime, her skull would be on display once it returned from its CAT scan.

More than 2000 visitors to the institute had already seen some of Sue's bones in Rex Hall, a small building behind the institute used for preparation and storage. "We didn't have a formal exhibit, but whenever tourists came through, I'd pull 'em off the display floor or out of the gift shop and take 'em to see Sue," says Larson. He admits that he was happy for any excuse to visit his prize find. "Even when I wasn't working on her, I'd go talk to her, fondle her bones. It was my daily ritual."

The institute's announcement of its museum plans energized Hill City. Although the town lay little more than a stone's throw from Mount Rushmore, it had little to offer tourists save an old gold mine and a short ride on an antiquated passenger train. Its economy had been stagnant for years. Average per-capita income in 1990 had been $9000, a full $3000 below the national poverty level. A museum featuring the world's greatest *Tyrannosaurus rex* would not only put the place on the map, it would put money in people's pockets.

It soon appeared that Sue might be keeping company with a fellow named Stan in the new museum. In April the institute announced the discovery of another *T. rex*. This specimen, about 60 percent complete, appeared to be gracile and was presumed to be a male. Larson named him after Stan Sacrison, the amateur paleontologist who found him on a private ranch in the badlands of Buffalo, South Dakota, about 90 miles west of Faith.

The institute's excavation team spent two weeks camped out at the site. Unfortunately, rain and snow dampened the ground, making removal of the bones difficult. While waiting for the weather to change, Larson and Wentz returned to Hill City and the preparation of Sue.

NASA had finally agreed to conduct the CAT scan of the skull. In mid-May, Larson and Wentz would be taking it to the Marshall Space

Flight Center in Huntsville, Alabama. They still had to finish preparing the skull and determine how best to pack it for the road trip south.

Three days after the institute made public the discovery of Stan, the April 29 *Rapid City Journal* broke a story that was, literally, news to Peter Larson. "Sioux Say 'Sue' Theirs," proclaimed a large, bold headline in the *Journal*. "The Cheyenne River Sioux say the world's largest *Tyrannosaurus rex* was illegally taken from land on their reservation in 1990," began reporter Bill Harlan's story. According to the article, the tribe asserted that Sue had been discovered within the boundaries of the reservation and that federal law prohibited taking such fossils without the tribe's permission. The Sioux had not filed a lawsuit, but months earlier they had passed a resolution calling for an investigation. "In Sioux Falls, U.S. Attorney Kevin Schieffer confirmed he had been investigating the case for months," the paper noted.

The story took everyone at the institute by surprise. Their only inkling that the tribe claimed ownership of Sue had come shortly after the find. At that time Maurice Williams had sent them a copy of the resolution with a note in which he wrote, "What an exercise in B.S., vindictiveness, jealousy, hatred, etc. You do what you think, but I would find it hard to even recognize their childish efforts." Larson chose to ignore the resolution.

Convinced that he had legally purchased the "theropod skeleton Sue" from Williams, Larson had also chosen to ignore a letter that the rancher had sent him on November 10, 1990, about three months after the discovery: "I didn't sell the fossil to you. I only allowed you to remove it and clean it and prepare it for sale." The institute had heard nothing from Williams or the tribe about Sue over the last 18 months.

Now, the *Journal* reported, Gregg Bourland, the chairman of the tribe, said that the Sioux had agreed to a partnership with Williams to determine the future of the dinosaur. That future might include making a cast of the *T. rex* for a museum on the reservation, the sale of a limited number of replicas of the skeleton, and donation of the bones to a college, said Bourland.

Larson, at the Stan site, was unavailable for comment, but the institute's attorney, Patrick Duffy, told the *Journal* that Williams's sale of Sue to the institute was legal. "I'm convinced that the tribe has no claim to the fossil at all," he said.

Over the next two weeks, Larson was more interested in getting Stan out of the ground and Sue's skull into a crate than he was in the tribe's claim, which he considered meritless. Wentz had extricated the skull from the pelvis by tunneling between the hips and the lower jaw bone. Once the other bones around the skull were removed, Larson and Wentz planned to get the skull into a wooden shipping crate that they had designed.

The crate was 6 feet long by 3 feet wide by 3 feet high; each of its six sides could be removed individually, allowing for better viewing or photography of the skull from various angles. The pair would cover each of these six sides with a removable layer of protective foam. The skull still sat on its original, though cut down, pallet. Larson and Wentz planned first to build four sides and the top of the crate around the skull—fastened to this pallet. They would then wrap the skull itself in industrial-strength plastic, turn the whole thing over, remove the pallet, and make a new bottom for the crate. NASA had told them that the skull should be as clean as possible; any plaster or foil would compromise the CAT scan.

As the skull's May 17 departure date neared, more details of the tribe's claim surfaced, as did pretenders to Sue's crown. The local *Timber Lake Topic* reported that only three days after the Larsons had first announced the discovery of Sue in September of 1990, the tribe had passed resolution E-335-90CR demanding return of the skeleton and requesting assistance from the Bureau of Indian Affairs. The tribe still asserted ownership, but Williams seemed to be distancing himself from the council. He told the *Topic* that there was no binding contract or sale. "Williams said he didn't know what the $5000 check was for but that 'maybe it might have been for mitigation,'" the paper reported.

It was unclear what Williams meant by "mitigation," but it was clear that he stood to benefit if the sale was invalid and the dinosaur was his. Gordon Walker, a private collector from British Columbia, told the *Topic* that he had offered Williams $1 million for the *T. rex* if he could get her back. He urged Larson to "come to his senses" and do right by Williams.

Who could declare the sale invalid and determine ownership? Schieffer said he was contemplating filing a complaint in federal court against the institute and that if he did, he would handle the case per-

sonally. Wait, said Mark Van Norman, an attorney for the tribe: A civil action in tribal court might be the best way of resolving the matter.

On May 12, five days after the *Topic* article appeared and five days before the scheduled departure to Huntsville, Larson received a call from his longtime friend Dr. Clayton Ray, a curator of vertebrate paleontology at the Smithsonian Museum of Natural History in Washington, D.C. Ray reported that someone identifying himself as an FBI agent had called the museum and asked how best to pack and move *T. rex* bones. Larson suspected that the caller had been Walker, trying to stir up trouble. Larson had recently angered the Canadian collector when he refused to verify that a "dinosaur egg" with an "embryo" in it that Walker was attempting to sell was genuine. "It wasn't a dinosaur egg. And the 'embryo' was a quartz formation," Larson says.

Larson told Duffy about the call. Duffy immediately phoned Schieffer. "We had the only *T. rex* around," Duffy later explained. "So I asked him if the government was planning to seize Sue. He told me no."

Schieffer says he didn't say no or yes. "I didn't divulge anything. I can't imagine any prosecutor confirming or denying anything where you are under investigation. It would be extraordinarily stupid to talk about it."

Who's telling the truth? Perhaps each man is. Upon receiving such a phone call, few U.S. attorneys would confirm that a seizure was imminent; to do so would give the subject of an investigation time to alter, destroy, or hide evidence. Says one former assistant U.S. attorney: "You generally say you can't comment on it. At the same time it's not unusual to, without lying, try and give the impression that nothing is being planned."

On May 13, Larson and Wentz continued readying Sue's skull. They were visited by National Park Service Ranger Stanley Robins, with whom Larson had served on the South Dakota State Historical Society Task Force on Paleontology. Robins frequently came by the institute to check on the preparation of the fossil. On this day he observed the progress they were making with the skull, bought some Sue memorabilia at the gift shop, and bade good-bye. At the time, Larson and Wentz didn't give the visit a second thought.

Larson went to bed that evening in a good mood. In 96 hours Sue would be on her way to Huntsville, where the technology of the 1990s would unlock mysteries 67 million years old. While there, Larson would talk to NASA about something else: "I wanted to volunteer to go to Mars and collect fossils," he says.

He slept well that night. As usual he had dinosaur dreams. It was the last good night's sleep he would get for a long time.

3

YOU BETTER GET
OUT HERE, PETE

"One million dollars."

Seated toward the back of the room, Stanford "Stan" Adelstein, a wealthy businessman from Rapid City, South Dakota, started to raise his auction paddle. He had journeyed east on behalf of the Black Hills Institute with $1.2 million in his war chest, while Peter Larson waited back in Hill City with butterflies in his stomach. "I thought just because of all the hype and the fact that she was a media star that she might bring as much as half a million or a million dollars," Larson would later say. "People were making these wild predictions, but they didn't have a clue of what the market was like. Because if you looked at the market, there was no way to predict this price. What had happened, of course, was the notoriety. The persona of Sue had created the story. I mean she was the most famous fossil in the world."

When Karl Häberlein put his *Archaeopteryx* up for sale in 1862, Sir Richard Owen was superintendent of the British Museum. Owen sent a representative to Bavaria with an offer: 500 pounds. Häberlein said no; the 700-pound price, like the specimen, was set in stone. Owen met with his trustees and persuaded them to give him an additional 200 pounds, and soon the British Museum had added another feather to its cap.

No doubt Owen, a superb anatomist, would have examined the specimen from head to toe—if part of the head hadn't been missing. As it was, he did conduct a thorough analysis of the fossil. Acknowledging that the *Archaeopteryx* did possess some reptilian features, Owen pointed to the feathers as "unequivocally" proving that the creature was a bird. There was no evidence that this was one of Darwin's missing links, he reported to the Royal Society in November of 1862.

Only a handful of *Archaeopteryx* have been discovered since Owen's report. One of the most significant of these was found in a museum, rather than in the field. In 1970, John Ostrom, a paleontologist at Yale University, "unearthed" the specimen at the Teyler Museum in Haarlem, the Netherlands. Apparently on finding it in 1855, von Meyer had misidentified it as a pterodactyl.

Ostrom found a lot of similarities between *Archaeopteryx* and dinosaurs that led him to conclude that the *Archaeopteryx* had evolved directly from theropods. Not everyone in the scientific community agreed. Birds are warm-blooded, while dinosaurs were cold-blooded, they argued. Ostrom admitted that he was uncertain whether dinosaurs were endothermic like birds and humans. But that didn't matter, he said. The anatomical similarities between *Archaeopteryx* and smaller carnivores, particularly coelurosaurian dinosaurs, demonstrated the link.

Bakker, who studied under Ostrom at Yale, was not the least bit uncertain: dinosaurs, at least the carnivores, were warm-blooded, he argued in 1975. Among his arguments: the ratio of dinosaur predators to prey is almost identical to the modern ratio of warm-blooded predators to prey—about one predator for every 10 prey. Cold-blooded predators like crocodiles, he noted, require far less food and live in densities where the ratio is the same in relation to prey.

Bakker is perhaps the best known paleontologist in the world. His long, bushy beard certainly makes him the most recognizable. His way with words—he has written novels as well as treatises—makes him the most quotable. And his fearlessness in forwarding new theories (along with his willingness to criticize his fellow PhDs) often makes him the most controversial. In addition to forwarding new theories on warm-bloodedness, he was the first to put feathers on dinosaurs, doing so as early as 1967 in his undergraduate thesis at Yale. He was also the first to propose a wholly new classification system that separates dinosaurs from other reptiles and puts them in a new class with birds.

Many scientists disputed Bakker's ideas about warm-bloodedness, trying to poke holes in the rationale for his predator-to-prey conclusions. Undaunted, he pointed to the dinosaur's skull as further evidence. Holes located near the brain's blood supply may have worked as radiators to cool the hot blood, he argued.

Because theropods and birds shared many characteristics, Larson, too, suspected that the *T. rex* was warm-blooded. "Sue may be more closely related to the hummingbird than to, say, the triceratops," he said. He was hopeful that NASA's CAT scan of Sue's skull would provide more light than heat in the warm-blooded vs. cold-blooded debate.

The trip to Huntsville should also answer questions about the brain size, if not the intelligence, of Sue and her fellow *T. rex*. Horner reported that the CAT scan of the skull of the Wankel/Museum of the Rockies' find revealed that the *T. rex* brain, while smaller than that of an elephant or rhinoceros, "was larger than that of almost all reptiles and other dinosaurs; proportionately, it was nearly as big as that of some birds." He added that "to be a bird brain is no compliment by our standards, but for most members of the animal kingdom, the comparison is flattering."

Dr. Dale Russell, curator of the North Carolina State Museum of Natural History, was one of many paleontologists interested in seeing what the CAT scan of Sue's skull would reveal about brain size. In 1981, Russell had speculated about the "evolution of intelligence" in dinosaurs if they had not become extinct. This speculation was informed by his study of *Troodon formosus,* a small bipedal carnivore from the Cretaceous period that he found in Alberta in 1968.

Working with Ron Sequin, a taxidermist, Russell created a fiberglass model of the "dinosauroid"—the twentieth-century version of what the dinosaur might have become if its evolution had not been interrupted. The model stood for years in the National Museum of Natural Sciences (now the Canadian Museum of Nature) in Ottawa, Canada, next to a reconstruction of the *Troodon*. It looks decidedly more like a human than a dinosaur. It stands upright on two legs and has a large, round skull, a short neck, and no tail.

Most of Russell's assumptions about the dinosauroid stemmed from the fact that *Troodon* had a brain weight to body weight ratio that compared favorably to that of early mammals. The paleontologist guessed, therefore, that the creature would have had a brain weight to body weight ratio similar to humans if it had survived the last 65 million years. If this assumption is correct, the skull would have had to be larger to accommodate the larger brain. This leads to the shorter neck theory; carrying this larger skull on a long, horizontal neck like that of *Troodon* would have been problematic. The upright body leads in turn to the absence of the tail, which would no longer have been necessary to counterbalance the body.

Like the dinosaur, the model features scaly skin, three-fingered hands, and no external sex organs. Unlike the dinosaur, it has a belly button; Russell speculated that the dinosauroid would have evolved from laying eggs to bearing its babies live. No nipples are discernible; just as birds feed their babies regurgitated food, so would the dinosauroid, Russell guessed.

Of course, Russell's speculation about the dinosaur of today would not be necessary if the world had not kissed all dinosaurs (and about 70 percent of all other species) good-bye some 65 million years ago. What caused this mass extinction? Scientists and lay people alike have forwarded a wide range of theories. In a classic 1964 article in the *American Scientist* titled "Riddles of the Terrible Lizard," Glenn L. Jepson, a Princeton scholar, catalogued the "What killed the dinosaurs?" theories, from the sublime to the ridiculous:

> Authors with varying competence have suggested that dinosaurs disappeared because the climate deteriorated (became suddenly or slowly too hot or cold or dry or wet), or that diet did (with too much food or not enough of such substances as fern oils; from poisons in water or plants or ingested minerals; by bankruptcy of calcium or other necessary elements). Other writers put the blame on disease, parasites, wars, anatomical or metabolic disorders (slipped vertebral discs, malfunction or imbalance of hormone and endocrine systems, dwindling brain and consequent stupidity, heat sterilization, effects of being

warm-blooded in the Mesozoic world), racial old age, evolutionary drift into senescent over specialization, changes in the pressure or the composition of the atmosphere, poison gases, volcanic dust, excessive oxygen from plants, meteorites, comets, gene pool drainage by little mammalian egg eaters, overkill capacity by predators, fluctuation of gravitational constants, development of psychotic suicidal factors, entropy, cosmic radiation, shift of Earth's rotational poles, floods, continental drift, extraction of the moon from the Pacific Basin, drainage of swamp and lake environments, sunspots, God's will, mountain building, raids by little green hunters in flying saucers, lack of standing room in Noah's ark, and paleoweltschmerz.

In 1980, Luis Alvarez, a Nobel Prize–winning physicist, and his son Walter, a geologist, offered their own theory after unusually high levels of the element iridium were found in rock dating back to the time of the mass extinction. Iridium, rarely found on earth, is more common in rocks from outer space. The Alvarezes concluded that an asteroid or comet as big as Mount Everest had struck earth while traveling 100,000 miles per hour. The impact, they argued, threw up a dust cloud that darkened the sky and caused temperatures to fall so low that most living creatures could not survive.

Much of the scientific community reacted to this theory with skepticism until a huge crater was found in Mexico's Yucatan Peninsula. Still, there exist almost as many theories today as in 1964. Bakker, for example, suggests that dinosaurs died off from disease not a cosmic calamity.

In *The Complete T. Rex*, Horner not only admits that he doesn't know what caused dinosaurs to become extinct, he confesses, "I don't really care. . . . I'm interested in how they lived." He adds, "Of course, to figure out how (they lived), I need the bones of dead dinosaurs." Like Sue's skull.

Peter Larson awoke early on May 14, 1992, to finish preparing that skull for its trip to Alabama. He had just climbed out of the shower in his trailer behind the institute when he heard Lynn Hochstafl, one of the institute's preparators, call his name. He looked at the clock. It was 7:30 AM.

"You'd better get out here, Pete. This place is crawling with FBI agents," Hochstafl said.

FBI

"Are you kidding?"

Assured that this was no joke, Larson dressed quickly and hurried outside. There he saw several men and women wearing blue jackets bearing the large yellow letters "FBI." These agents were surrounding the institute with yellow police ribbon, with the warning "SHERIFF'S LINE DO NOT CROSS" printed in bold black letters.

Larson moved inside the building to his office, where he was met by several members of his confused and terrified staff . . . and dozens of FBI agents and Sheriff's officers. "All told there were approximately 35 law enforcement officers, all with guns and sour looks," Larson recalls.

Park Service Ranger Robins was also present. "What are you doing here, Stan?" Larson asked. Robins said he was part of the government's investigative team. "You mean you were spying on us?" Larson asked. His voice betrayed his astonishment. Robins, who has since died, looked away, says Larson.

Two agents, William Asbury and Charles Draper, approached Larson and handed him a federal search warrant alleging criminal activity, including felonies of stealing from government land and from tribal land as well as violations of the Antiquities Act of 1906. U.S. District Judge Richard Battey had issued the warrant on the basis of affidavits sworn by Asbury and Robins. It demanded that the institute surrender

> all the fossil remains of the *Tyrannosaurus rex* skeleton (commonly referred to as "Sue") . . . and other fossil specimens . . . taken from an excavation site on the property of Maurice Williams . . . (and) all papers, diaries, notes, photographs, including slides, memoranda, tape recordings, videotapes, maps, butcher paper, and samples or other records relating to the excavation of the *Tyrannosaurus rex* ("Sue") and other fossil specimens.

"This can be real easy or real hard depending on whether or not you are willing to cooperate," Larson recalls Agent Asbury saying.

The usually calm Republican paleontologist was not quite ready to cooperate. "Shaking, I asked them to indulge me for a few minutes, to please close their ears if what I was about to say offended them. I then proceeded to read them the Riot Act," he writes in *The Rex Files.*

"How can you come into our place of business and seize our property without due process?" he fumed. "Don't you know the worst thing you could possibly to do Sue is move her? Please leave her where she is. I'm a law-abiding citizen. I'll give you whatever guarantees you need, but please don't move and damage her." Then he called his lawyer.

By the time Duffy had arrived from Rapid City, virtually everyone in Hill City knew that the FBI had raided the institute and was preparing to haul Sue away. A sizable portion of the population in this part of the country is naturally suspicious of the federal government. "[Its] Gestapo attitude doesn't go down well in a small community like this," one citizen told the *Journal*.

Beyond this philosophical disdain for federal intervention in the lives of its citizens, the residents of Hill City had a pragmatic reason for wanting to keep Sue at the institute. The *T. rex* promised to make them prosperous by attracting tourists. In the weeks since the formal announcement of the Black Hills Museum of Natural History, the town was on the verge of enacting a "bed and booze" tax, with part of the sales tax increase going to purchase the 10 acres overlooking town where the institute proposed to build the museum.

Unwilling to see their dreams shattered, the local residents quickly mobilized. They first asked area loggers to bring in their logging trucks to block the convoy of government vehicles parked outside the institute. After deciding that such a confrontation would be dangerous as well as futile, they resorted to more traditional forms of protest. Bette Matkins, the town's former mayor, hastily scribbled a sign reading, "PLEASE LEAVE OUR CITY IMMEDIATELY WITHOUT SUE." She was joined in front of the institute by approximately 50 townsfolk carrying signs and shouting the same sentiment. Soon children poured out of the Hill City elementary school and ran down Main Street to see what was happening to their beloved Sue. Some of the protesters, young and old, were crying.

The FBI ignored the tears and the chants. Duffy arrived to find them going through the institute's files. "No corner was safe from these people. We felt violated," says Larson. Several paleontologists and students from Larson's alma mater, the South Dakota School of Mines and Technology, were also on hand. Their job: to supervise the safe loading of the fossils in question.

Larson and Duffy pleaded with Agent Draper to allow the CAT scan of Sue's skull to proceed. NASA, Bakker, and editors from *National Geographic* magazine (which planned to photograph the scan for an upcoming article on Sue) echoed this plea. They all argued that the raid was jeopardizing the safety of one of the most important finds in paleontological history and delaying important research. "Mr. Draper seemed sympathetic to our plight and spent a great deal of time on the telephone talking to the powers that be in an attempt to allow the CAT scanning to take place," remembers Larson.

The primary "power that be" was U.S. Attorney Schieffer. Technically, Schieffer was the "acting" U.S. attorney. He had been sworn in with that designation about five months earlier in December 1991. The appointment by South Dakota's Republican U.S. Senator Larry Pressler had surprised many in the legal community because Schieffer had no federal trial experience. Rather, he had served as a legislative aide to Pressler since 1982. He had attended Georgetown University's law school in Washington, D.C., at night and earned his degree in 1987.

Schieffer had not initiated the investigation of the institute's excavation of Sue and other fossils. The office he took over had been investigating fossil collecting in South Dakota for about a year, says Assistant U.S. Attorney Bob Mandel. Mandel, who headed and continues to head the Rapid City office, says that in the weeks before Schieffer's arrival, there had been talk of initiating a sting or sending in an operative to expose illegal collections and sales.

The institute had actually come to the attention of federal authorities in the mid-1980s, thanks to Vincent Santucci, a young graduate student in vertebrate paleontology. In 1985 and 1986, Santucci had worked part time as a ranger at Badlands National Park in South Dakota. During his first days on the job, he was shocked to see people engaged in what he calls "shopping behavior"—the taking of fossils from park lands. "I thought everyone knew you don't do that," he recalls.

Eventually, Santucci caught an old-timer who freely admitted that he had been making a living for 25 years taking fossils and selling them to rock shops and collectors and leading fossil hunters on gathering expeditions in the park. The man showed Santucci a scrapbook that included sales receipts. The Black Hills Institute was among those purchasing the illegally obtained fossils, says Santucci.

Santucci says the old-timer explained that if he didn't take the bones they would erode away and be lost forever. "He told me he was rescuing the fossils," says Santucci, who is now the chief ranger at Fossil Butte, a small national monument in southwest Wyoming. Nevertheless, the Park Service brought charges against him. He was found guilty and fined $75. The paucity of the penalty outraged Santucci. "It was a life-changing experience," he says. He vowed to devote his career to the enforcement of laws preventing the theft of treasures he considered every bit as important and valuable as archaeological artifacts.

In 1986 Santucci began looking into the activities of the institute. He recorded his suspicions in reports to the park service. When Stan Robins began working at Badlands National Park, he read the reports and talked to Santucci, who had moved on to the Petrified Forest. Together, the two rangers tried to devise a plan for exposing the institute and others whom they felt might be engaged in illegal collecting. They believed that an arrest of a collector with a high profile might deter others from stealing fossils from public lands. "We wanted to demonstrate in a public way that we can't accomplish our mission when people deliberately take fossils from public lands—when they know it's illegal and do it for economic gain. We can't protect the public, [if] we don't have legislative or regulatory tools in place," says Santucci. Although the U.S. attorney's office won't confirm it, it appears that Robins had brought the institute to the attention of the FBI and prosecutors before Sue was unearthed.

Santucci learned of Sue's discovery before the institute even made the find public. He was alerted by a commercial collector who had visited the institute shortly after the bones were brought back to Hill City. Having been unable to build a strong enough case against the Larsons to date, Santucci's eyes now lit up. Sue was a "charismatic *T. rex*," explains the ranger. Surely the media, the public, and the fossil-collecting world would have to take note if a "white man had ripped off a Native American" by taking such a valuable specimen from Indian lands. But were they Indian lands? Santucci called Robins, and the investigation was under way.

Robins served on a state committee looking into fossil collecting, as did Larson. So, too, did representatives of Native American reservations in South Dakota. Soon after Robins heard about Sue from Santucci, the

Cheyenne River Sioux tribe lodged its complaint with the U.S. attorney's office.

"In my first week as the U.S. attorney, I was asked to look into [that complaint]," Schieffer later explained. "I did and just approached it from an objective viewpoint and went by the law." That viewpoint led him to authorize the raid.

On the night before the raid, Schieffer sent out press releases and called a press conference for the morning of May 14 in Rapid City. As a result, several television film crews came to Hill City as the FBI moved through the institute. So, too, did Schieffer, a fit-looking, well-dressed man in his late thirties with short black hair and a neatly trimmed mustache. Late in the morning, he met with the press outside the institute. Then he headed inside.

Larson and Duffy again pleaded their case. "Let the specimen stay where it is and we'll make any guarantees you demand. Above all, please let the CAT scan proceed as planned," Larson remembers saying. Drue Vitter, the newly elected mayor of Hill City, made the same request. Schieffer turned them all down. The skull was evidence in a criminal investigation at that point, Schieffer explains. "Letting it go (to NASA) would have been a nightmare from a chain of custody standpoint."

"Did you see what I saw?" Duffy asked Larson as they walked away from Schieffer.

"You mean the television makeup he [Schieffer] smeared all over his face?"

To Larson this was evidence that Schieffer had an ulterior motive: Perhaps he planned to run for political office in the future and felt that this case would give him the necessary exposure with the electorate. Knowing the television cameras would be at the institute, he must have wanted to look his best, Larson reasoned. Schieffer would later admit that he was wearing pancake makeup, but he said he had put it on for an earlier, unrelated television interview. He dismissed Larson's claim that seizure of the fossil was to be his stepping stone to elected office. Raiding the institute was so unpopular that it could only hurt someone with political aspirations, he said.

He adds that he had not even intended to go to the institute on the day of the raid. He did so only at the behest of the beleaguered FBI agents who did not want to deal with the media. "It was them or me," he says. "So I bit the bullet." He suspects that the institute or Duffy wanted him there in hopes of creating a spectacle for the press. He adds that when Duffy had first called him to ask if a raid was being planned, "I didn't know him from Adam," but others in the office did. "All the old hands assumed this was gonna be a media event" when they heard that Duffy represented the institute, he says.

The institute staff, NASA, and the residents of Hill City were not the only ones surprised by the government's raid. Cheyenne River Sioux chairman Bourland said that he was shocked that the federal government had confiscated the bones. "The Cheyenne River Sioux tribe is not the enemy. We didn't have anything to do with sending these guys in . . ."

Bourland and Larson and their respective supporters were just as surprised by Schieffer's pronouncement on the day of the seizure: "The fossil is property of the United States. Period." Not the property of the institute. Nor of the tribe. Nor of Maurice Williams. Schieffer explained that shortly after buying the land from a white homesteader in 1969, Williams had exercised his right to have the property held in trust for 25 years by the U.S. Department of the Interior under the Indian Reorganization Act passed in 1934. Among other advantages, the act allowed Native Americans to avoid paying taxes on the land they put in trust. Putting one's land in trust, as Williams had, limited one's ability to sell the land to a non-Indian without the permission of the department. Williams had neither sought nor received permission from the department.

Schieffer did not address the question of whether selling dinosaur bones was the equivalent of selling land. In fact, the controlling law he cited suggested just the opposite. The federal Antiquities Act of 1906 prohibited the removal of fossils from any land "owned or controlled" by the United States without a permit, he said. This included trust land, he continued, adding that the institute did not have a permit. Schieffer explained that other federal statutes made it "a crime to steal government property and to take property off Indian lands."

The invocation of the Antiquities Act raised eyebrows in some legal circles. Congress's immediate concern in passing the act had been the protection of Native American sites in the Southwest. According to Schieffer, the act "effectively reserved to the United States all rights in objects of antiquity by mandating that their excavation be 'for the benefit of . . . recognized scientific or educational institutions, with a view to increasing knowledge of such objects, and . . . for permanent preservation in public places.'"

Did a fossil qualify as "land" held in trust? Or was a fossil an "object of antiquity?" Apparently Schieffer wanted the courts to sort this out.

When? Schieffer said he didn't expect any arrests soon, reported the *Journal*. But he added: "This isn't exactly jaywalking. This is a priceless antiquity."

As Schieffer spoke, the FBI continued its search of institute files. "They took everything," Larson says. "Even the letters I had written to schoolchildren."

Meanwhile, the crew from the School of Mines was rewrapping Sue in plaster jackets similar to those she had worn 21 months earlier. Larson and Wentz received permission to continue preparing Sue's skull so that it could be safely loaded for its trip to its "holding cell" at the School of Mines. (The pair would later speculate that Schieffer waited to send in the FBI until Ranger Robins reported that work on the skull was almost done; otherwise it could not have been hauled away.)

At the time of the raid, Larson was hosting a friend from Canada. Leon Kinsbergen, originally from Holland, was a Holocaust survivor who had lost his wife and children in the concentration camps. "He was terribly upset during the seizure [of Sue]," says Larson. "I tried to comfort him by explaining, 'Don't worry, Leon. This is America.' Leon replied, 'That is what makes it so terrible. I can't believe this is happening in America.'"

Neither could the people of Hill City. Over the next three days, as Sue was loaded into boxes, hoisted onto flatbed trailers, and hauled away by the National Guard, 200 men, women, and children joined in protest outside the institute. Their homemade signs reflected their feelings. "WHAM, BAM, THANKS UNCLE SAM." "SHAME ON YOU." "FBI IS U.S. GESTAPO."

By the time the last government truck carrying Sue was loaded, Larson and his fellow workers were in tears. So, too, were some sympathetic members of the National Guard. Larson noted that it had taken 17 days to dig Sue from the cliff, thousands of hours had been invested in her preparation, and countless more hours had been spent thinking about every aspect of her care. And then: "They packed, loaded, and moved the finest dinosaur ever found in three days."

Larson did more than wring his hands during those three days. Besides working with Wentz to prepare Sue's skull for removal, he worked with Mayor Vitter and tribal chairman Bourland to broker a deal that might quickly resolve the matter. On Friday, May 15, day two of the seizure, the three men met in Hill City. "A common concern that the fossil might be put in a place of the federal government's choosing helped bring us together," Bourland told the press after the meeting. "Let's not turn this into a fight between the U.S. government and Hill City and the Cheyenne River Sioux tribe. Sue's big enough for everybody."

Perhaps. But Schieffer seemed to be holding all the cards, er, bones. He called the meeting a positive step but reasserted that he had no doubts that Sue was the property of the federal government. "My primary directive under the Antiquities Act is to bring these [objects] back under the public domain," he said.

By the time Larson, Bourland, and Vitter met for a second time in Rapid City on May 20, everyone from politicians to physicists to paleontologists had weighed in on the matter. South Dakota U.S. Senator Tom Daschle scheduled a town meeting in Hill City for May 25. Marshall Center physicist Ron Beshears, who was to have supervised the CAT scan in Huntsville, expressed disappointment. "We look at a lot of rocket nozzles, and a lot of hardware components," he said. "This [Sue] is an interesting project. We sure hope they will get this thing settled."

Meanwhile, Currie, Bakker, and Horner publicly expressed anger over the government's action. "I'm reminded of the last scene in *Raiders of the Lost Ark,* in which the U.S. government, having seized the Ark of the Covenant, locks it away in a forgotten warehouse," said Bakker.

Horner agreed. The seizure was "absolutely ridiculous," he told *The New York Times*. He explained that federal officials could have left Sue where she was until the legal dispute was resolved. "Who's going to walk off with a tyrannosaur?" he asked.

Prosecutor Mandel had a different take. "There's nothing unusual about seizing evidence that is the subject of a criminal investigation," he would say later. "It's not our way to trust putative defendants to take care of evidence for us."

Schieffer says other options were considered. "We struggled with it. The only other thing we could do—which might have been better from a public relations standpoint—would have been patently unfair [to the institute] if we were wrong—lock the doors of the business. In retrospect that might have been more media savvy, but we do things by the book instead of the camera."

In the same front-page article in which Horner was quoted, the *Times*'s science reporter Malcolm Browne touched on an issue that would grow in importance over the coming months: the conflict between commercial fossil hunters like the Larsons and the academic community. Some scientists associated with universities or museums resented the commercial hunters, arguing that they had no qualms about selling important fossils to private collectors, thereby removing them from the realm of scientific study.

Did the Larsons fit this profile? "Many prominent paleontologists say the Black Hills institute takes great care to preserve scientific assets and make them available to scholars," wrote Browne. Bakker was one of these paleontologists. He admitted that his original inclination was to lump the Larsons with the bad guys. "I thought, Oh, these guys are just ripping out bones and selling them to Japan," he said. "Then I saw their lab." He told Browne that the Larsons weren't only fossil hunters, they were "scientists . . . who kept meticulous scientific records of their finds."

The *Times* wasn't the only media giant to find the seizure newsworthy. *Newsweek,* the *Wall Street Journal,* and *The Times* of London also carried feature articles, and NBC's *Today Show* did a segment from Hill City that included footage of local seventh graders presenting Larson with $47.50 for "Sue's Freedom Fund."

By this time the protest had extended to Rapid City. On Monday, May 18, about 30 people marched outside the federal building that housed the U.S. attorney's office. "G-MEN CATCH *CRIMINALS* NOT DINOSAURS," read one sign.

On the previous day the *Rapid City Journal*'s editorial page had featured a large political cartoon of a "G-man" dressed as a caveman and carrying a club that read "FBI." This figure stood outside a building identified as the institute. "He says he's here to see 'Sue,'" said another figure, who was poking his head out from inside the institute.

Three days later, on May 20, the *Journal* ran an editorial highly critical of Schieffer's actions. Titled, "Was an Army Necessary?" it read in part:

. . . Poor Sue. It all seems so unnecessary. Instead of being allowed to remain at the institute, where trained and experienced workers could care for her and where she eventually would be shared with the public and with inquiring scientific minds, she has been exposed to potentially damaging transport and has been boxed, sealed, and put in storage. . . .

It appears doubtful that Schieffer's heavy-handed action was required to resolve Sue's fate. The parties involved—with the possible exception of Schieffer—are willing to . . . attempt to work out an agreement.

It seems entirely possible that a plan can be developed to ensure that Sue is on public display, that she is available for science to learn as much as possible about her and her species, and that she doesn't become some entrepreneur's cash cow. And perhaps Hill City can have its museum, and the tribe can benefit.

Accomplishing those goals is possible. And it very likely was possible without armed government intervention.

When Peter Larson left Hill City later that morning for his meeting with Mayor Vitter and tribal chairman Bourland, he agreed with the *Journal* that a plan was indeed possible. After the FBI had left, some townspeople had taken scissors to the police tape that had surrounded the institute. Driving down Main Street, Larson passed a row of utility poles wrapped in yellow ribbon. Perhaps those ribbons would soon be down and Sue would be back where she belonged, he hoped.

At the first meeting, Bourland had voiced the tribe's desire for a cast of Sue for display on the reservation in Eagle Butte. Like the Larsons, he envisioned Sue as the centerpiece for a larger facility—one that, in the tribe's case, would include materials gathered during the Black Hills land claims of the 1800s. Those materials were currently in a vault in Sioux Falls because there was no money to catalog and display them. Presumably, the cast of Sue would draw tourists to the reservation. "It would only be right that there's a cast of Sue at Eagle Butte," Larson had said on May 15. As Eagle Butte was 200 miles from Hill City, the Sioux's Sue and the institute's Sue would not be in direct competition.

Although the tribe and the institute appeared to be heading toward an agreement, they were forgetting (or ignoring) one very important fact. Schieffer had said that neither of them owned Sue. So how could they make an arrangement to determine the fossil's future?

In an interview with *USA Today* published on May 20, Schieffer himself seemed to suggest that such an arrangement might be possible, and he outlined the conditions. These conditions appeared consistent with Larson's intentions from the day he had announced the discovery of Sue. Said Schieffer: "The institute must guarantee that [the fossil] will be displayed permanently in a public museum and would not be sold or traded out of the country or into a private collection."

The meeting later that day between Bourland, Larson, and a delegation from Hill City including Mayor Vitter lasted four hours. The participants did not immediately reveal the substance of the discussion to the public. Later, however, Bourland told the *Journal*'s Bill Harlan that the tribe had proposed that Sue be displayed in a new museum in Hill City. In return, ownership of the *T. rex* would be transferred to the Sioux. The tribe would then give Hill City a 99-year lease on the dinosaur and would help Hill City build the museum. Some Indian artifacts might also be displayed there. The tribe would share in some of the museum's revenues.

Larson told Harlan that he had rejected the proposal. The institute still claimed ownership, he explained. Despite their inability to reach an understanding, the parties agreed to meet again, this time in Eagle Butte.

That meeting never took place. On May 21, one day after the Rapid City talks, Bourland announced that there would be no more negotiations. He claimed that Larson had violated a "gentlemen's agreement" by issuing a press release earlier in the day. In the release, the institute had said that it was "accepting the terms set forth yesterday by U.S. Attorney Kevin Schieffer . . . in *USA Today*."

Bourland interpreted this as an attempt to cut a deal with the government behind the tribe's back. He added that Schieffer had termed the press release a "cruel hoax" and had told him that he had rejected the institute's offer.

Duffy said that he had talked to Schieffer after the article had appeared and that the prosecutor had neither accepted nor rejected the proposal. He added that nothing in the press release violated the "gentlemen's agreement," whose purpose was to keep tribe–institute negotiations confidential.

Larson tried to strike a conciliatory note. "Boy, I'm really sorry they're mad at us," he said. "This was not done to insult the tribe, and if it did, I apologize."

The institute's attempt to deal directly with Schieffer may have lacked tact, but it did make sense from a tactical point of view. For the moment anyway, Schieffer had the fossil and the power to decide what should be done with it. In the *USA Today* article, he had appeared to publicly address the institute, explaining what they had to do to get Sue back. The institute's press release was an attempt to let Schieffer know, publicly, that they were willing to do what he wanted.

But the institute's promise to put Sue in a museum and never sell her out of the country or to a private collector was not enough to persuade the U.S. attorney to return the fossil or to drop his investigation. Ownership of the fossil may have been the subject of debate at that point, acknowledges Schieffer. Federal laws could be interpreted to suggest that the tribe and Maurice Williams had legitimate claims to Sue. But, Schieffer adds, it was clear to the government from the beginning that the institute had no claim to the bones. (This doesn't explain, however, why he suggested a possible solution to *USA Today* involving the institute.)

Dropping the investigation was not a possibility either. "The media never grasped that the criminal investigation had precious little to do with Sue," he says. Before Sue had even been discovered, the U.S. attorney's office was looking into allegations that the institute stole fossils from public lands. The Sioux's complaint about the theft of the dinosaur just happened to coincide with that inquiry.

Once they realized that Schieffer was not going to return Sue voluntarily, Larson and Duffy weighed all their options. In doing so, they recognized that although Sue had lain undisturbed for 67 million years, time was now of the essence.

Sue was being stored inside a 40-foot steel tank in a machine shop next to a boiler room at the School of Mines and Technology. Larson worried that the bones would be irreparably damaged if they sat there much longer. Temperature variations in the room could lead the bones to expand and contract and eventually split. And such variations were inevitable; the shop was not air-conditioned. The fossil faced another danger, according to Larson. Sue had pyrite in her bones. Pyrite is iron sulfide, an inherently unstable material that decomposes. When iron sulfide combines with water from the air, sulfuric acid is created, which slowly dissolves bone. As the bone and pyrite dissolve, calcium sulfate is created. Calcium sulfate is in gypsum, a fast-growing mineral that would actually push the bone fragments apart. Temperature variations like those in the machine shop would accelerate the process.

How could the splitting and dissolving and fragmenting be prevented? By spreading out the bones so that they could be watched, Larson believed. By controlling the temperature so that it did not vary more than a degree or two each day. By continuing to clean, prepare, and glue the bones so that fractures and cracks could be addressed as they occurred. In short, by returning Sue to the controlled conditions and experienced preparators at the institute.

On May 22, eight days after the raid, Duffy went to court. The headline in the *Journal* was an editor's dream: "Institute to Sue for Sue." *Tyrannosaurus rex* had become Tyrannosaurus lex.

4

TAKING A HOWITZER TO A FLY

"One million five hundred thousand," said Redden.

Stan Adelstein had raised his paddle to make the bid, which was $300,000 more than he had intended to spend. Why not? he said to himself. If he could get Sue for one point five, the bank would surely give him the extra money.

In 1770, 200 years before Sue was discovered, workers in a chalk quarry in Maestricht, the Netherlands, found a pair of fossil jaws more than 3 feet long. They summoned a local fossil collector, a retired German military surgeon. He promptly took the extraordinary specimen.

Anatomists called in to determine the jaws' origins were puzzled. One thought that they may have come from an ancient whale. Another speculated that they belonged to a huge marine lizard. This seemed impossible. No lizards this big had ever been spotted in the water or on land. "Still," Wilford writes, "as so many Europeans came to suspect, this did seem to be a prehistoric monster, something that might have lived before Noah, possibly before Adam, and passed out of existence. But this was, it seemed, something they were not sure they believed in."

The find received much notoriety in Europe. Despite his inability to identify the jaws, the surgeon proudly displayed them. In time, he was sued by the man who owned the land on which the jaws had been found. In all likelihood this was the first lawsuit ever fought over a fossil. The court held that the jaws belonged to the landowner. Eventually, they were displayed in a glass shrine in the residence of the local canon. But the story does not end here. The fossil would go on to contribute to a revolutionary scientific theory forwarded in 1801 by the Paris-based paleontologist and anatomist Georges Cuvier.

It took 25 years and the French army to get the specimen to Paris. In 1795, while the army of revolutionary France was fighting in Holland, its leader, General Charles Pichegru, received a most unusual order: Seize the famous jaws of Maestricht. Legend has it that Pichegru offered 600 bottles of wine to the man who could liberate the fossil. Whether by wine, song, or some other means, Pichegru accomplished his mission. He shipped the jaws home, where they were soon examined by Cuvier, the foremost authority on marine life and fossils of his day.

Cuvier concluded that the jaws had indeed come from a huge marine lizard that no longer existed and must have lived long, long ago. This analysis, coupled with his analysis of ancient elephant bones unearthed in Paris, led him to proffer the theory of extinction. The fossils clearly suggested that some forms of prehistoric life no longer existed, he argued.

While the concept of extinction may seem perfectly obvious as the twenty-first century begins, it was heretical to many in the early days of the nineteenth century. As Wilford notes, the Bible suggested something quite different. Ecclesiastes 3:14 reads: "I know that, whatsoever God doeth, it shall be forever: Nothing shall be put to it, nor any thing taken from it." Still, Cuvier and the jaws of what would eventually be named *Mosasaurus* carried the day.

As the custody battle for Sue demonstrated, mankind's (and government's) instinct to possess rare fossils was as alive in 1992 as it had been two centuries earlier. The question of who owned Sue had become so tangled that the institute felt it necessary to name four parties as defendants in its lawsuit: the United States Department of Justice, the Department of the Interior, the Cheyenne River Sioux tribe, and the South Dakota School of Mines and Technology, where the *T. rex* was being stored and, where rumor had it, she might be displayed. The institute did not sue Maurice Williams, as he did not claim ownership of the fossil.

In its suit before Judge Battey, the institute claimed ownership and sought to "quiet title" under the Quiet Title Act, a federal statute that, with certain exceptions, allows a party to sue the United States in a civil action to adjudicate a disputed title to real property. But the name of the act was just about the only thing surrounding Sue that was quiet. One day after filing the lawsuit, the institute showed questionable taste at the Dakota Days parade in Rapid City. Its float was a flatbed truck with crates like those used to transport Sue after the seizure. On the truck, a mock FBI agent admonished the crowd to stay back lest they be wrestled to the ground.

The following day, it was Schieffer's turn to demonstrate questionable judgment. In an op-ed piece in the Sunday *Journal*, he discussed the pending criminal action. "Because legal action is being considered, I cannot discuss many specific facts of this case or argue its legal merits. But it is appropriate to explain the policy basis for actions taken." The policy explanation that followed looked remarkably similar to an argument of the legal merits.

The institute and the U.S. attorney weren't the only ones making noise. Enter the normally staid Society of Vertebrate Paleontology: "The Society of Vertebrate Paleontology firmly supports the action of the U.S. attorney's office in Rapid City as regards the siezure [sic] of a specimen of *Tyrannosaurus rex* that apparently was collected on federal lands without proper permitting procedures as required under federal statutes," began a one-page press release issued May 20 on SVP letterhead.

The release concluded: "The [SVP] is extremely concerned as to the heightened activities of commercial collectors in recent years, resulting in loss of invaluable, nonrenewable paleontological specimens to foreign interests. . . ." It was signed by Dr. Michael O. Woodburne, professor of geology and vertebrate paleontology at the University of California, Riverside. Woodburne, a past president of SVP, was currently the chairman of its Government Liaison Committee.

The last line of the document read: "This statement has been authorized by the president of the Society of Vertebrate Paleontology." This apparently came as news to the president, Dr. C. S. Churcher, a professor of geology at the University of Toronto. Reached by the *Journal*'s Harlan, Churcher said that he had not seen the press release before Woodburne had issued it to the news media. "I don't think we should make a judgment here," said Churcher. He agreed that no fossil should be taken illegally from any jurisdiction, but he said that he did not understand the U.S. law on the subject. The seizure of Sue had, he said, caught him by surprise. "From a Canadian point of view, I find it utterly ridiculous," the SVP president added.

Seven months earlier Larson had been warmly received at the SVP convention in San Diego when he had spoken of Sue's "startling surprises" and had invited fellow scientists to come to Hill City to help study the *T. rex*. He had belonged to the SVP since 1974. "As a member of the society, I'm deeply appalled by Michael Woodburne's actions," he told the *Journal*. Appalled but not surprised. Andrew Leitch once told *Discover* that Woodburne and a few like-minded colleagues were conducting a "witch hunt" against commercial collectors and considered Peter Larson "the antichrist."

Some at the institute suspected that a handful of SVP members jealous of Larson's success may have pressured Schieffer or his superiors in Washington, D.C., to seize the dinosaur by asserting that Sue's skull was about to be sold to a private party. Larson has never found a smoking gun, but he feels somewhat certain that a member or members spread the story that Sue's skull wasn't really going to NASA—that, instead, it was going to the Georgia-based defense contractor, Martin Marietta Corporation (which subsequently merged with Lockheed to become Lockheed Martin). In reality, Martin Marietta was merely helping to ship the skull to Huntsville. Schieffer insists that the timing of the raid had nothing to do with the imminence of the CAT scan; the investigation had merely reached the stage where the seizure of evidence was appropriate.

Newspaper editorials and commentaries in several cities across the country asked why Sue had to be seized at all. The *Cleveland Plain Dealer,* for example, acknowledged that "an international bidding war for fine dinosaur fossils has made tough stewardship of the U.S. fossil record essential," but, the paper continued, " . . . Schieffer displays tough stewardship only of his media image. . . . The recent action targets a respected research institute that has been at the forefront of dinosaur studies instead of the fossil brigands who plunder and destroy key evidence of our 65-million-year-old dinosaur and geologic past." Writing in the *Washington Times,* conservative columnist Bruce Fein lamented: "The dispute epitomizes criminal justice madness and the potential for societal strangulation by omnipresent intervention. . . . Doesn't the *Tyrannosaurus rex* farce suggest that the legal tipping point has been reached or passed in the United States?"

Native American writers also weighed in. In his "Notes from Indian Country" column in the *Lakota Times,* Tim Giago did not pass judgment on the legal issues. Focusing on the raid itself, he observed: "By charging into Hill City as if storming the beaches of Anzio, the federal marshals put the Cheyenne River Sioux tribe and other Indian tribes in a bad light. The near-violent actions made it appear the tribes had something to do with it when, in fact, the entire scenario developed in the office of . . . Kevin Schieffer. Certainly the bones weren't going to jump up and run off on their own."

The *Rapid City Journal*, having already made its feelings known, asked its readers what they thought: "Who should get Sue?" Only 5.4 percent of the more than 1000 respondents to the "informal and wholly unscientific poll" voted for the federal government. A handful more, 5.6 percent, said Sue belonged to Maurice Williams. Thirteen per cent said the tribe was the rightful owner. And an overwhelming majority, 76 percent, said Sue should go home to the institute.

The institute certainly wanted her home—and sooner rather than later. Five days after filing the lawsuit requesting title to Sue, Duffy, a dapper, handsome man in his middle thirties, with neatly coiffed brown hair, returned to court. This time he filed a motion for injunctive relief. Arguing that the fossil was being irreparably damaged in the machine shop, the motion sought the immediate return of Sue to the institute pending Judge Battey's decision of the ownership question.

In affidavits attached to the motion, both Peter and Neal Larson noted several "physical and biological processes" that threatened the *T. rex*. The physical processes included mineralization, hydration, crystallization, and oxidation. Biological processes included mold and mildew forming in the cracks of the bone, conditions that could quickly erode the bone surface. Some of these problems had been caused by failing to let the wet plaster casts "protecting" Sue's bones dry properly and by using wet toilet paper to help seal the bones in the plaster. The brothers said that the casts should be removed and preservation of the bones should continue. The institute, they claimed, was the only venue nearby where such work could be done and where Sue could be properly protected.

Judge Battey, a former prosecutor appointed to the bench by Ronald Reagan, rendered a decision the following day. The judge appeared to the robe born. He was a tall, distinguished looking gray-haired man in his sixties partial to bow ties. His "presence and charisma" reminded Peter Larson of Charlton Heston.

In making this decision, Battey never addressed the allegations claiming damage to the fossil. Instead, he viewed the action for injunctive relief as an attempt to regain evidence seized in a criminal investigation. Unwilling to permit this, he denied the motion. Duffy immediately appealed the decision to the United States Court of Appeals for the Eighth Circuit.

In his brief to the court, Duffy noted that more than mold and mildew threatened Sue. The bones were precariously stacked and were positioned dangerously close to a boiler and corrosive chemicals. He also made it clear that the government could have access to Sue, should the court order her returned to the institute. On June 26, the court ruled. "This case concerns the care and custodianship of a 65-million-year-old pile of bones named Sue," began Judge Frank Magill in his opinion for the three-judge panel. He then addressed Sue's current living arrangement: "The federal government has stored this irreplaceable relic under circumstances that even its own experts describe as inadequate."

Turning his attention to the raid itself, Magill wrote: "The government has admitted it does not need 10 tons of bones for evidence in its criminal investigation. . . . We find the government's rationale for the seizure inadequate. The seizure not only keeps [the institute] from accessing the fossil, but it deprives the public and the scientific community from viewing and studying this rare find."

Despite making findings so favorable to the institute, the court did not order the government to return Sue to Hill City. Instead, it ordered Judge Battey to hold a full hearing to establish proper custodianship pending determination of the ownership question. Battey complied, setting the hearing for July 9.

Schieffer had trouble with the Eighth Circuit's ruling. He noted that the court had seemed to diminish the severity of the potential charges facing the institute. Magill had written that the seizure was "based on an investigation into criminal charges that could result in, at most, 90 days in jail and a $500 fine." Wrong, said the acting U.S. attorney. In fact, the search warrant for Sue had listed two misdemeanors and one felony, and the maximum penalty for removing antiquities from federal land was six months in jail and a $5000 fine for individuals or a $10,000 fine for companies.

Was the felony still under consideration? Schieffer refused to comment when asked this question by a Rapid City radio station after the Eighth Circuit ruled. But one thing was clear: A federal grand jury had been convened on June 16 to consider criminal indictments against the Larsons and others.

Five days before that grand jury had convened, Schieffer had slapped the institute with a subpoena for virtually all its business records—some 50,000 documents, letters, maps, videotapes, and photographs, Peter Larson estimated. Duffy immediately went to court to quash the subpoena, saying that it was far too broad. "They came and asked for every record for them to comb through and will try to find a crime, and we believe that's excessive," he said. Judge Battey allowed the subpoena to stand.

Neal Larson was angry. "[The subpoena] has nothing to do with the court case. They're just trying to keep us busy, up in the air, and keep us mad," he told the press. "And it's working."

Larson wasn't the only person angered by the action. Four days later, Wentz, who had gone from chief preparator of Sue to chief organizer of the "Free Sue" movement, led some 40 protesters to Pierre, the state capitol. The delegation lined up outside the federal building and held up signs reading, "DINOSAURS ARE FOR CHILDREN, NOT PRISON," "SEIZE DRUGS, NOT FOSSILS," and "HONK FOR SUE."

Of course, neither the shouting nor the honking had any impact on the criminal investigation. Within a week of being convened, the grand jury was visited by a forgotten face. "Fossil Finder Testifies in Sue Case" ran the headline in the Sioux Falls *Argus Leader* on June 21.

Since parting company with Peter Larson almost two years earlier, Sue Hendrickson had lived up to her "Indiana Jones" nickname. She had continued searching for amber in the Dominican Republic and Mexico. She had also spent much of her time searching for shipwrecks in the waters off Cuba and the Philippines with a team led by famous French marine archaeologist Franck Goddio.

The Cuban effort was particularly noteworthy. Until Goddio's arrival in Havana in 1991, Fidel Castro had refused to let outsiders look for the treasure-laden galleons that had sunk on their way back to Spain. Hendrickson was largely responsible for Castro's turnaround. During a sailing regatta in 1979, she had become friends with Vicente de la Guardia, the director of Carisub, Cuba's state-run marine archaeology effort. She had sent him the latest literature on diving techniques and equipment for almost ten years before seeing him again in Cuba in 1988. Two years later he had asked her to recommend a team to help Carisub. She had suggested Goddio, who then enlisted her to dive and serve as liaison with the government. She was the only woman on his diving crew.

Although they were no longer dating, Hendrickson and Larson had kept in close contact during this period. They saw each other at paleontology society meetings and talked frequently on the phone. "We were still best friends," says Hendrickson.

When the *T. rex* had been seized, Larson had phoned Hendrickson in Seattle. She wasn't home. She returned the call from France a few days later. "I kept saying, 'Really?' and laughing," she remembers, "but then it started sinking in—that it wasn't a joke. At that point everyone at Black Hills was in shock, but they were thinking, This is too bizarre. It will all get straightened out."

Shortly before her grand jury appearance, Hendrickson had joined Larson at the Tokyo Fossil and Mineral Show, where she helped at the institute's booth. She had only been in Japan for a short time when she received a call from her mother in Seattle. FBI agents had phoned Mrs. Hendrickson and told her they had a subpoena for her daughter. The government wanted Sue to testify before the Rapid City grand jury investigating the institute. Mrs. Hendrickson had referred the agents to her son, John, Sue's brother, who was a partner at a major Seattle law firm.

On her return to Seattle, Hendrickson met with her brother and other attorneys from his firm before calling the FBI. The lawyers advised her that the government often applies pressure on potential witnesses by threatening to prosecute them. Therefore, she should try to strike a deal for full immunity from prosecution before testifying before the grand jury. Although she felt she had done nothing wrong, Hendrickson agreed to this plan. The lawyers had received some oral commitments from the prosecution before Hendrickson journeyed to Rapid City, but nothing regarding immunity had been put in writing.

In Rapid City, Hendrickson was represented by Gary Colbath, a lawyer who had previously served as an assistant U.S. attorney. For three days, Hendrickson and Colbath met with Schieffer, Assistant U.S. Attorney David Zuercher, and FBI agents. As the prosecutors debriefed his client, Colbath attempted to finalize the immunity deal. Despite his one-time membership in the fraternity of prosecutors, Colbath, who died in 1999, had little kind to say about the conduct of Schieffer or Zuercher: "Kevin Schieffer had no trial experience, and David Zuercher was just vindictive."

The government offered and withdrew immunity to Hendrickson three times, all the while threatening to bring charges against her. Finally, a deal was struck, and Hendrickson testified before the grand jury for two days. "The prosecutors thought I was going to be the star witness against Pete—the ex-girlfriend who spills her guts," she says. "Then they realized I didn't fit that mold and that I didn't have any information to give them anyway. I didn't know of anything the institute had done wrong."

Hendrickson recalls that as the appearance dragged on, Zuercher grew increasingly frustrated with her. "He'd show me pictures of some site and ask if I could identify it and recall collecting there. The grand jury was totally bored. Zuercher would get so upset with my answers that he'd have me leave the room until he could calm down."

The prosecutors told Hendrickson that she was not to talk to the press and not to have any contact with anyone from the institute. "Can they do this to me?" she asked Colbath. Aware of the fragility of the arrangement with the government, Colbath advised her to follow these instructions. "That's when the nightmare really began," Hendrickson recalls. "These were my closest friends and we couldn't talk during the period we really needed each other. I can remember driving through South Dakota thinking I should just stop at the institute or call, but I couldn't." She followed the case through her friend Casey Carmody, a Seattle mineral and fossil dealer who had friends at the institute.

Although Hendrickson couldn't talk to the media, Colbath could. "I don't ever recall such a Mickey Mouse procedure or being jerked around as much as I was appearing with Ms. Hendrickson," he told the press. "She has confirmed what everyone has said, that no one at the institute did anything unlawful or intended to circumvent the law. . . . The U.S. attorney's office was taking a howitzer to a fly, so I wouldn't be surprised to see these people severely charged. The federal government has never been able to say, 'We screwed up.'" Seven years later Colbath felt the same way. He termed the investigation one of the worst abuses of prosecutorial power that he had seen in his 25 years of criminal defense work.

While the grand jury considered whether to indict the Larsons criminally, the SVP did indict them professionally. On June 26, the same day the Eighth Circuit ordered Judge Battey to hold a hearing to determine temporary custody of Sue, the SVP executive committee sent Schieffer a letter stating its "strong objection to the removal of the tyrannosaur 'Sue' from its present location . . . to the Black Hills Institute, a commercial fossil-collecting business." Writing on behalf of the committee, SVP Secretary-Treasurer Dr. Robert Hunt argued that "professional paleontological expertise" was available at the School of Mines and Technology to properly care for and preserve the fossil. He also cited "a recently prepared conservator's report [that] attests to the fact that the skeleton is not deteriorating in any way."

The following day, 13 SVP members, including Hunt, Woodburne, and President Churcher, got personal. In a second letter to Schieffer, they wrote:

> Officers of the Black Hills Institute, a commercial fossil business, have not, in our opinion, published scientific studies demonstrating serious scientific expertise in dinosaur research in the past and have not demonstrated an ability to carry out noncommercial scientific work, nor do we believe they are able to identify the best experts to undertake such a study. In fact, it is our opinion that the "institute" often chooses associates who support their views on commercial collecting, whereby some fossils are sold, others are retained for their own purposes which often reflect the long-term commercial interests of the business and not those of the scientific discipline of vertebrate paleontology or the general public.

Larson had no doubt that Woodburne and Hunt, longtime critics of the institute were behind the letters. Hunt, curator of vertebrate paleontology and professor of geology at the University of Nebraska, seemed to admit as much in a 1992 interview with the *American Lawyer*. He told the magazine that the signatories to the letters felt that "the press was portraying the professional community as divided on this issue," when in reality paleontologists were "unanimous in condemnation of this supposed institute."

On this particular issue, Hunt's own research proved suspect. The professional community of which he spoke seemed far from unanimous. After this letter was sent to Schieffer (and the executive committee recommended expulsion of any member found collecting on federal lands), Clayton Ray of the Smithsonian, an SVP vice president slated to become president in three months, resigned in protest; he had been an SVP member for three decades. Dr. Donald Wolberg, a paleontologist at the New Mexico School of Mines and acting president and secretary of The Paleontological Society, also resigned. He had been an SVP member for 26 years.

Bakker termed the two letters "scandalous, quite probably libelous." The notion that the institute chose only associates who supported its view was particularly ridiculous, he said; the Larsons invited everyone to Hill City, including Hunt. As for the institute's scientific skills, he said, "We stiff PhDs, speaking in Latin, wearing elbow pads, criticize the Larsons for not having degrees, but their research is better than ours."

Not surprisingly, at the hearing to determine temporary custody of Sue, Wolberg and Bakker testified on behalf of the institute, while Hunt testified as an expert for the government. Their respective evaluations of Sue's condition either demonstrated that scientists sometimes reasonably differ in their analyses of similar data or that scientists sometimes unreasonably allow personalities and politics to influence their decisions. Or both.

In its opinion calling for the hearing, the Eighth Circuit had written: "If the parties could come to a common understanding on this issue, the court as well as the public at large would be well served." As in the past, the parties couldn't reach an understanding. And so 100 people packed Judge Battey's courtroom on July 9. While most of those in attendance supported the institute, the government was not without friends. The Cheyenne River Sioux tribe, whom the institute had voluntarily dismissed from the lawsuit, had filed an amicus curiae (friend of the court) brief supporting the U.S. position that Sue should temporarily remain at the School of Mines.

Wolberg took the witness stand first. During his testimony, he compared Sue to "an original or a second draft of the Constitution" and a "Van Gogh or Whistler or Remington." This *T. rex*, he said, was a "totally unique specimen, educationally, culturally, scientifically."

Wolberg had visited the machine shop at the School of Mines. What did he think of the present storage conditions of this unique specimen? Duffy asked. "Trying not to overreact, I was appalled," said Wolberg. High heat and humidity, the possibility of fire and flood, and the presence of corrosive chemicals all threatened Sue. So, too, did the fact that she was being stored in the Rapid City flood plain. She was already suffering from pyrite disease and her plaster jackets were cracking and in need of repair. "It would have been the last place I would have put this."

Where did Sue belong? "I would give it back to the Black Hills Institute on the quickest train you can find. . . . They can appraise, treat, curate, and store it if needed," Wolberg said.

Cross-examined by Schieffer, Wolberg admitted that Peter Larson was a close friend. Was he testifying as a friend or an objective scientist? Both, said Wolberg.

Leitch and Bakker agreed with Wolberg's findings. Bakker, too, saw evidence of pyrite disease. It was important that the fossil be cleaned and examined frequently, he said. The institute was the only place to store Sue during the many months that would pass before ownership was decided. (Prior to the hearing, Judge Battey estimated that it could take up to two years before the ownership question was resolved.)

The government also presented three experts. Pat Leiggi, a preparator for the Museum of the Rockies, testified that Sue should remain where she was. "More damage will occur by moving the fossil," he explained. Sally Shelton, a natural history conservator for the Texas Memorial Museum at the University of Texas in Austin agreed. She said she found little difference between the housing at the institute and that at the School of Mines. Transporting Sue back to Hill City was 100 times more likely to damage the fossil than allowing it to remain at its present location, she said, likening the effect of moving the bones to the effect of setting them afire.

At the government's request, Shelton had performed an x-ray test on a small piece of bone to check for pyrite. She stated that the risk factor of pyrite disease was "very low." She added that monitoring equipment should be installed in and around the container housing Sue and that the bones should be checked regularly. The School of Mines had already begun installing such equipment.

Bakker, who had respected Shelton's previous scholarship, was perplexed by her findings. "She just looked at a chip from the corner of one bone. That's like looking at one chip of a mansion and concluding it doesn't have termites," he says.

Hunt, however, seconded Shelton's opinion. "From what I've seen . . . [Sue] is in better condition than most people realize. . . . The building is adequate to protect this fossil," he testified, adding that the level of pyrite in bones was so low that they might not even require treatment. How many bones had he examined? Under cross-examination by Duffy, Hunt said he had looked at only three bones and analyzed only one small fragment for pyrite.

Through Hunt, Schieffer introduced into evidence the two letters written by SVP members critical of the institute's scientific capabilities and supportive of the current storage plan. As Larson's lawyers noted, however, the signatories had never seen the institute. Or Sue. Or the machine shop.

The institute and the government lawyers weren't the only ones asking questions of witnesses. Examined directly by Judge Battey, Peter Larson said he would spend whatever it took to preserve Sue if she were returned to the institute. Larson also promised to post a bond on Sue guaranteeing that the institute would not use her for any commercial purposes.

Neal Larson also took the stand. The questions Schieffer asked him had little to do with Sue's physical fitness and much to do about the institute's moral fitness to house her. Larson admitted that after receiving Ray's phone call about the FBI's call to the Smithsonian, he had changed the dates and descriptions on several boxes of fossils. He explained that he had hoped that this would prevent their seizure in the event of a raid. "I was scared and stupid," he said. "I don't know why I did it." He emphasized that when the agents arrived, he immediately told them what he had done and corrected the dates.

Court recessed for the weekend after two days of testimony. That Saturday, July 11, the long-awaited Black Hills Museum of Natural History had its grand opening. The opening did not take place on the 10-acre site where the Larsons hoped to build the museum; the not-for-profit corporation they had established had nowhere near the money to purchase the land, and, of course, the status of the main attraction was still in limbo. The several hundred people attending the event instead gathered at the museum's "temporary home," the institute. The building's facade now featured a huge painting of the *T. rex* crying behind prison bars. "Free Sue," read the caption.

Mayor Vitter, Bakker, and Wolberg spoke to the gathering, as did Gus Hercules, the Libertarian party candidate for the United States Senate. Hercules had been one of the most persistent critics of the raid since day one, endearing himself to the Larsons and others who thought that in seizing Sue, the government had once again gone too far. Over the last two months, the longtime Republican paleontologists and many of their friends had evolved into libertarians.

When testimony concluded two days later, Judge Battey immediately announced his decision: Sue would stay where she was. "The court has been presented with no credible evidence that the fossil is suffering damage," he said. He added that both Hunt and Shelton were particularly strong witnesses. In contrast, he wrote in his formal opinion three days later, "The court finds that the interests of the plaintiffs [the institute] and their expert [witnesses] are more commercial than scientific. Their interest in science is secondary to the financial value of the fossil's possession." Larson found this last statement particularly perplexing. The judge had apparently ignored the fact that long before the seizure, Larson had announced that Sue was not for sale and had invited SVP members to study her.

Having settled the temporary custody issue, Battey turned to the terms of Sue's interim storage. The judge said he would allow only necessary care and maintenance of the fossil until ownership was decided. Both the institute and the government could submit requests for storage conditions to him by August 15. Finally, the judge had harsh words for Neal Larson. Changing the dates on the boxes "may well constitute obstruction of justice," he said.

The Larsons and their supporters were devastated. Judge Magill's Eighth Circuit opinion had suggested that Sue was inadequately housed in the machine shop and was better off at the institute. Now Judge Battey had ruled otherwise, ordering that he would permit no "curation of this specimen" until he determined ownership. How was science served when research was halted? they asked. As the courtroom emptied, Neal Larson sat head in hands, weeping. Duffy announced that once again he would appeal Judge Battey's decision.

At one point during the hearing, Duffy had attempted to put the dispute in perspective for Judge Battey: "No federal courtroom should end up the tool of what is essentially a faculty war between paleontologists," he argued. Like many wars, this one seems to have been fought for control of turf deemed valuable by the combatants. In this case the battle was over public lands rich with fossils of scientific and commercial import.

The law with respect to the taking of archaeological remains from federal lands has been clear since the enactment of the Archaeological Resources Protection Act of 1979 (ARPA). The act imposes stiff penalties for the removal of human artifacts and skeletons from public lands. It has this to say regarding fossils: "Nonfossilized and fossilized paleontological specimens, or any portion or piece thereof, shall not be considered archaeological resources . . . unless found in an archaeological context."

Similar legislation or clarity regarding paleontological resources found on public lands does not exist. Schieffer's assertion that the Antiquities Act applied to the removal of fossils seemed to fly in the face of several court cases and Department of Interior memoranda from 1977 and 1986. In the absence of one controlling federal statute, a patchwork permit system had evolved. Before collecting or removing any specimens, fossil hunters were supposed to get the permission of whichever federal agency was responsible for the land in which they were interested, typically the Bureau of Land Management or the U.S. Forest Service. With limited resources for policing, these agencies enforced the permit system sparingly and randomly.

In the mid-1980s, a panel of the National Academy of Sciences addressed the question of fossil collecting on public lands. The panel solicited input from land managers, professional groups, amateurs, and commercial collectors. Larson was a member of the panel. He noted that there were thousands of amateur collectors in America, a few hundred university-affiliated paleontologists, and about 100 commercial collectors. He also noted that every year millions of fossils "weathered out" on federal land so that they were visible. Sadly, they soon eroded and were lost to everyone because no one was there to collect them. He concluded that if more people were allowed to look for fossils, more fossils would be found.

In 1987, after three years of study, the panel issued its findings. The 243-page report recommended that all public lands in the United States with the exception of national parks and Department of Defense property be opened "for scientific purposes" to commercial collectors and amateurs as well as paleontologists affiliated with museums and universities. "Paleontology is best served by unimpeded access to public lands," the panel declared.

The panel also recommended that "fossils of scientific significance should be deposited in institutions where there are established research and educational programs in paleontology." But this did not appease hard-line opponents of universal access. The war between usually civil scientists was on.

The brethren of the SVP were divided on this issue. However, a vocal faction of SVP members, including Hunt and Woodburne, lobbied against adoption of the recommendation. As a result, it died.

Woodburne turned down a request to be interviewed for this book after receiving a list of questions concerning commercial collecting in general and the Larsons in particular. He did, however, issue a statement summarizing his position:

> The full educational and scientific potential of all vertebrate fossils can only be realized if they are collected in a manner consistent with those goals, and retained in perpetuity in the public domain in educational and scientific institutions that, under permit, steward those collections for all Americans. Commercialization of fossil vertebrates is inconsistent with those goals.

When it was suggested to Hunt that this book would explore both sides of the debate over commercial collecting, he responded that journalists have "had the wool pulled over their eyes" by "glib" proponents of the universal access camp. To suggest that there is a debate suggests that the viewpoint of the commercial collectors has some credence, he said, when in reality it has no credence. He said that his own statements are often taken out of context by writers and that he generally prefers not to participate in the dialogue.

Larson was willing to discuss what he believes to be the origin and history of Hunt's animosity towards him. He points to an incident that occurred on Hunt's turf in the mid-1980s. The institute had gone to Nebraska in search of horses from the Pliocene Epoch 5 million to 2.6 million years ago. Larson asked a rancher near the town of Culbertson for permission to look on his property. The rancher informed Larson that a University of Nebraska team had previously dug on the land, but they hadn't been back for some time. Larson then contacted the graduate student who had supervised the dig and asked if the institute could explore the site. This student, who was in Hunt's program, "said he was finished with his research and had no problem with our collecting," according to Larson. The institute then found three-toed Pliocene horses. "Hunt was furious," says Larson, who adds that, to his knowledge, the University of Nebraska team has never returned to the site.

After this incident, Larson and Hunt clashed over political turf—SVP policy. When Hunt became chairman of the SVP's government liaison committee—the body responsible for formulating legislative proposals—he closed meetings to all SVP members except those on the committee, says Larson. Larson was allowed to speak to the committee, but he had to leave the room when finished. He protested this, as well as a poll about collecting that Hunt sent to those in the SVP. Believing the poll asked the wrong questions and that the data in the questionnaire about collecting were skewed, Larson sent his own questionnaire to members.

How did Hunt respond to Larson after these clashes? "He never screamed at me, but he treated me as if I weren't human," says Larson.

From his previous statements, it appears that Hunt truly believes in his position and is tenacious in advancing it. Commercial collecting, particularly collecting on public lands, is, to him, an evil. "This is a religion to him," says Larson.

Those who held Hunt's views had been pressing for a crackdown on commercial collectors before the seizure of Sue. The federal government had responded in the mid-1980s by undertaking an investigation of fossil hunters in Wyoming as well as in South Dakota. The institute had bought some specimens from the Wyoming hunters, and the government had subpoenaed records of these transactions. "We complied, gave them the documents they wanted," says Larson.

The government's claim that Sue had been discovered on federal lands seemed to energize the anticommercial faction. In addition to intervening in the lawsuit, the SVP's executive committee approved a resolution calling for prohibition of commercial collectors from public lands and authorizing the expulsion of SVP members who "engage in commercial collection and sale of vertebrates from federal lands." The committee's June 26 letter to Schieffer explained its reasoning:

> Commercial collecting of fossil vertebrates has promoted the charging of collecting fees on private lands. Professional vertebrate paleontologists cannot afford these fees for both financial and ethical reasons. As a result, professional paleontologists in the employ of universities and museums must rely on federal lands for access to important paleontological resources to continue their research programs. Hence, prohibition of commercial collecting on federal lands is essential to scientific progress in vertebrate paleontology.

During the summer of Sue, spurred by the renewed furor over commercial collecting on federal lands, U.S. Senator Max Baucus of Montana introduced legislation intended to give fossils the same protection that archaeological remains enjoy under ARPA. Baucus's proposal, titled "Vertebrate Paleontological Resources Protection Act," required that any collector—professional or amateur—secure a permit before picking up a vertebrate fossil and could collect specimens only if affiliated with a museum or school that "has no direct or indirect affiliation with a commercial venture" that collected fossils. "If we continue to allow these public resources to be sold to the highest bidder, we stand to lose crucial sources of scientific research and public education," Baucus said when introducing the bill. Violations of the act could be punishable by a $10,000 fine and up to one year in prison. A second offense could result in up to five years in prison and a fine of up to $100,000. *The New York Times*'s Malcolm Browne noted: "'Under the proposed law even a camper who wandered into a national park or Indian reservation and removed a fossil bone might be sent to jail for up to one year."

The Baucus bill eventually died in a Senate committee. But Baucus and the SVP executive committee did have a legitimate bone to pick with at least some commercial collectors. Clearly, there were some "fossil brigands." The Baucus caucus pointed to a study of some 40 fossil sites on public lands. Fully one-fourth of these sites had been vandalized, surveyors reported. The Larsons and other established commercial hunters who deal regularly with museums and universities as well as private collectors had no patience for these plunderers, either. They gave a bad name to the profession and ruined potential collecting sites and should be punished, Peter Larson said.

Larson, however, took issue with the SVP executive committees' jeremiad. A line-by-line reading did reveal several fallacies.

Responding to the charge that commercial hunters caused private landowners to levy collecting fees, Larson said, let he who is without blame cast the first stone (or fossil). He referred his critics to the period after the Civil War. At that time the search for dinosaurs moved from the eastern United States to the West, where Joseph Leidy had first identified the remains of *Trachodon* and *Deinodon* in 1858. Two scholarly paleontologists, Othniel C. Marsh, a professor at Yale and curator of the university's Peabody Museum, and Edward Drinker Cope, a former Haverford College professor who often conducted surveys for the United States—paid local landowners and "scouts" for bones and directions to fossil beds. The practice led Leidy to complain: "Professors Marsh and Cope, with their long purses, offer money for what used to come to me for nothing, and in that respect I cannot compete with them."

Larson and others also took issue with the SVP's assertion that financial and ethical considerations made it impossible for paleontologists in the employ of universities and museums to pay private collecting fees. First of all, not all private landowners charged collecting fees. Moreover, many of the universities and museums that employed paleontologists had no financial or ethical qualms about paying commercial collectors *directly* for their finds. Indeed, most of the Black Hills Institute's sales were to universities and museums.

It was unclear that these supposedly financially strapped and ethically bound university and museum paleontologists were truly limited to collecting on federal lands. But even if this were the case, why was, as the Hunt letter stated, "prohibition of commercial collecting on federal lands . . . essential to scientific progress in vertebrate paleontology"? There were millions of acres of federal land rich in fossils. Wasn't there enough land to go around for the commercial collectors as well as for the university and museum paleontologists? One was tempted to ask why Hunt and his colleagues, who thought so little of the scientific expertise of the Larsons and their ilk, wouldn't be able to demonstrate that in the field the fittest scientists survive.

Baucus and the SVP 13 had a point that, lamentably, some commercial collectors were finding important specimens and removing them from scientific inquiry (in the United States, at least) by selling them out of the country or to individuals instead of institutions. The Museum of the Rockies' Leiggi, who had encouraged Baucus to draft legislation, noted that a Japanese corporation had offered $4 million for the skull of the *T. rex* his museum unearthed in 1990.

Legitimate commercial fossil hunters did not deny that the willingness of deep-pocketed foreign museums and private collectors at home and abroad had raised the price that not-for-profit institutions in the United States had to pay for specimens. They noted, however, that commercial collectors had found many of the most important fossils that had ended up in the hands of American scientists. The institute had found Sue, the best *T. rex* ever, on what the government claimed was federal land. Was the SVP saying that because the Larsons were commercial collectors, science would have been better off if the dinosaur had never been found at all?

Bakker, for one, saw the SVP position as disturbingly elitist. "It's a classic ivory tower response," he told *Discover* magazine shortly after the seizure. "Because these people have their PhDs, they think they have some God-given duty to protect antiquities and fossils. They're like self-appointed guardians of the faith; they want to make fossils off limits to anyone without a doctorate. It's especially tragic because it threatens good amateurs—who've done more for the science than anyone."

In a 1998 interview, Dr. Louis Jacobs, a professor of geological sciences at Southern Methodist University and president of SVP, unashamedly admitted that paleontologists at not-for-profit institutions *should* feel superior to commercial hunters. "They [commercial hunters] don't play any role in science. . . . I think you could find any graduate student and they would have greater interest, greater dedication, greater desire to build a better world [than do commercial hunters]." These students and their professors are "dedicated to doing something good" as opposed to making money, Jacobs insisted. He seemed unimpressed with the argument that those at universities and museums might be moved by factors other than altruism, such as the opportunity for financial gain or professional advancement and recognition.

Larson bristles at Jacobs's blanket condemnation, then offers his own generalization about his critics. "There's always a small segment of any profession, a portion of the scientific community for whatever reason, that has a distrust. It's a feeling that some people have of inadequacy, envy—although they don't see it as that. They have a frustration at not really making any significant advances of their own. Then they see someone with no [formal] training or specific skills [like a Sue Hendrickson]. They feel they put in the early effort, they had the schooling, and they feel rewards should come to them.

"But there is more to it than schooling. You can't just stand at the foot of a mountain. You have to be willing to climb the mountain. A lot of armchair paleontologists have never grasped the fact that you have to work to discover something. If you want to find something in the field, you have to go out into the field. You can make wonderful discoveries opening museum drawers, but it's not the same thing."

Larson acknowledges that there isn't a lot of government money for paleontological research. "That is frustrating. But many of these [academics] earn comfortable livings. They could spend some of their own money to go out into the field. They have the time—but they don't do that."

As the war of words over the pros and cons of commercial collecting intensified, the battle for temporary custody of Sue moved toward a resolution. Almost two years to the day that Hendrickson had first spied Sue, the Eighth Circuit agreed to hear the institute's appeal of Judge Battey's decision to keep the bones at the School of Mines until the question of ownership was decided. The court would hear oral arguments from the parties on October 14 in St. Paul.

At the same time, the institute's original suit to determine ownership (not temporary custody) took an interesting twist. On July 31, the institute amended its original complaint that had sought to quiet title. In this new complaint, the institute abandoned the quiet title theory and did not request that the court determine ownership at all. Instead, the amended complaint raised a single issue: whether the institute's *claim* to Sue, based on its purchase from Maurice Williams, was superior to the government's need for the bones as evidence in a criminal action the government had yet to file.

Why would the institute abandon its claim that it owned Sue and argue now only that it had a superior possessory claim? Quite simply, the lawyers now felt that this was the best way to regain the bones. Judge Magill's June 26 opinion had, after all, stated that the "rationale for the seizure [was] inadequate." The opinion had also seemed to suggest that the government had to prove its need for the dinosaur, or give her back until the ownership question was finally settled.

The institute lawyers felt that if they could get Sue back temporarily under this theory, they would never have to give her up again. Their reasoning went like this: Both the tribe and the government were claiming ownership. The tribe was threatening to bring its own action for the *T. rex* in tribal court, where it might receive a more favorable hearing. But if it did so, it would have to name the U.S. government as a defendant. Invoking the defense of "sovereign immunity" (immunity from the jurisdiction of the court by virtue of its status as a government), the United States would never come to court, and the tribe's ownership claim could not be adjudicated. Similarly, if the United States sued for title to Sue in federal court, the tribe could also invoke sovereign immunity (or so Judge Magill suggested). And with the tribe absent, the federal court could not adjudicate its claim. Either the tribe or the government could waive the defense of sovereign immunity, but the institute lawyers did not think they would. And if they didn't, then Sue would remain at the institute permanently by default.

Whether or not the bones were needed by the government, the criminal investigation was proceeding . . . and expanding. Court documents filed by Schieffer alleged that the case involved reservation land and "other public lands" in South Dakota. The acting U.S. attorney also said that he was investigating "ongoing, multistate criminal activity." The government was obviously looking into more than the collection of Sue. No doubt the second subpoena requesting virtually all of the institute's business records was aimed at getting information about numerous other dealings.

By the end of August, Larson was beginning to learn which dealings. The government had subpoenaed documents from Nippon Express, an air freight company in Minneapolis. Nippon had once shipped a replica of a dinosaur back from Japan to the institute.

The names of grand jury witnesses were also surfacing. Subpoenaed by Schieffer, Clayton Ray had testified for two days. Why? During the raid, Larson had casually mentioned to an FBI agent that a friend from the Smithsonian (Ray) had called a few days earlier about a rumor that the bureau wanted to know how to move a dinosaur. Larson remembered that the agent had told him that the call may have constituted obstruction of justice.

Schieffer also subpoenaed Eddie Cole, a Utah fossil hunter with whom the Larsons had done business for 15 years. Cole told the *Journal*'s Harlan that Ranger Robins and FBI agent Asbury had visited him at his home. They were interested in a prehistoric turtle he had sold Larson in 1991. According to Cole, the pair insisted that he had taken the turtle from an excavated hole on Bureau of Land Management (BLM) grasslands. He said that the investigators knew that he and his family had camped near the hole; they even knew he had left 21 Merit cigarette butts there. They had videotapes and photographs of the site as well.

Cole denied the allegations. He said he had found the turtle on private land and bought it from a rancher. He had then sold it to Peter Larson, who helped with the excavation.

Cole was also questioned about a partial skull that he had sold the institute. He admitted that it may have come from state or federal land, but insisted he hadn't revealed that to Larson because "he wouldn't have bought it." Questioned by Harlan, Larson confirmed this. He insisted that he never knowingly bought specimens collected on public land, but acknowledged that he didn't always personally check the location. "We buy from people we trust," he explained.

The skull in question was that of a mosasaur.

Cole claimed that Robins and Draper had harassed and threatened him during their initial interrogation. Larson also told Harlan that he feared that the government was subjecting many of his clients and suppliers to the same treatment. Duffy added: "[The government] is sending word out planet-wide to the paleontology community that my clients are under investigation." He accused Schieffer of conducting a "scorched earth" investigation. "It's a frightening glimpse of justice under the New Order," he said.

On October 10, scores of institute supporters gathered at the Heart of the Hills Convenience Store Exxon & Super 8 Motel parking lot next to the institute for the "First Annual Sue Freedom Run, Walk, Hop, Skip, Jump, or Crawl." For a $10 fee that went to the Free Sue fund, participants received a *T. rex* T-shirt, post-race refreshments, and the chance to win prizes donated by more than 25 local businesses, including Hill City Jewelers, Jake's Casino at the Midnight Spa, Frosty's Drive-in, and Andrea's Chain Saw Sales and Service.

Four days later, institute lawyers gathered at the federal building in St. Paul for oral arguments on their appeal of Judge Battey's decision to let Sue remain at the School of Mines. They knew that this proceeding might be their best chance to win the *T. rex* her freedom. In June, Judge Magill's three-judge panel had seemed considerably more sympathetic than Judge Battey had ever been.

The institute and the government had previously submitted lengthy briefs with numerous cases supporting their respective positions. The court also had the 600-page record of the July hearing presided over by Battey, including the testimony of each side's experts on pyrite disease and dinosaur storage. Duffy felt confident that the briefs and the record favored his side, and he made the legal arguments as any lawyer would. Then he told the court how he really thought the case should be decided: "It comes down to this," he said, pausing for effect before continuing. "Who loves this dinosaur more?"

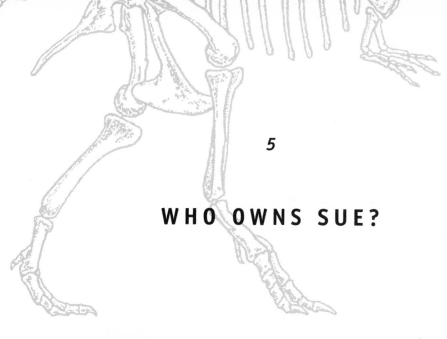

5

WHO OWNS SUE?

"Two million, three hundred thousand," Redden said. Stan Adelstein's paddle and Peter Larson's heart sank at the same time. "I guess we lost her," Larson sighed.

No moment of silence was observed. The bidding continued. "Two million seven hundred thousand." Larson couldn't believe the numbers.

Few newspapers outside South Dakota covered Sue Hendrickson's discovery of the greatest of the great dinosaurs. In contrast, the FBI raid on the institute was featured on several network news shows and garnered front-page headlines across the country and around the world. The seizure, not the science, was the story. Few reports focused on Peter Larson's findings about the anatomy and social habits of *T. rex*. Larson was much more compelling as a David fighting the Goliath of government than he was as a paleontologist.

This focus on personality rather than papers is hardly new. In April 1875, the National Academy of Sciences met in Washington, D.C. At the meeting, Yale's Othniel C. Marsh—the second person in history to be named a professor of paleontology, the first in America—presented important new findings on the development of the brain. The popular press yawned.

Reporters had gone to that gathering of the country's most respected scientists looking for something quite different. The *New York Tribune* wrote: "Had Prof. Cope been present we might have hoped for a battle of bones between him and Prof. Marsh, and possibly for an episode that would have served for a supplement to the meeting. . . . But everything was decorous and slightly dull." One scientific theory holds that, like clockwork, every 27 million years some monumental natural catastrophe causes mass extinction. The extinction of decorum and dullness in the world of science occurs considerably more regularly. The uncivil but highly entertaining late twentieth-century battle of bones involving Larson, the SVP, the federal government, and Native Americans reminded many who hunt stories about fossil hunters of the equally contentious but engrossing late nineteenth-century battle of bones involving Marsh, Cope, academic societies, the federal government, and Native Americans.

Cope vs. Marsh was not a lawsuit but a feud played out in the court of public opinion for almost 30 years. The hatred these two paleontologists felt for each other eventually eroded their images as men of character. At the same time, however, the rivalry moved them to almost literally move mountains and make many of the best fossil discoveries in history.

It apparently began with a mistake. In 1869, Cope restored the remains of a giant reptile, the *Elasmosaurus*, in the Museum of the Philadelphia Academy of Natural Sciences. Marsh, who had previously paid homage to their friendship by naming a new species *Mosasaurus copeanus*, came to see the display. His subsequent recollection of the events sheds insight into each man's personality:

> When Professor Cope showed it to me and explained its peculiarities I noticed that the articulations of the vertebrae were reversed and suggested to him gently that he had the wrong end foremost. His indignation was great, and he asserted in strong language that he had studied the animal for many months and ought to at least know one end from the other.
>
> It seems he did not, for Professor (Joseph) Leidy in his quiet way took the last vertebra of the tail, as Cope had placed it, and found it to be the atlas and axis, with the occipital condyle of the skull in position. This single observation of America's most distinguished comparative anatomist, whom Cope has wronged grievously in name and fame, was a demonstration that could not be questioned, and when I informed Professor Cope of it his

wounded vanity received a shock from which it has never recovered, and he has since been my bitter enemy. Professor Cope had actually placed the head on the end of the tail in all his restorations, but now his new order was not only extinct, but extinguished.

In his comprehensive biography of Marsh and Cope, *The Dinosaur Hunters,* Robert Plate paints pictures of two very different personalities with the very same goal. The methodical Marsh and the impetuous Cope were each consumed with being the most famous, most highly regarded scientist of the day. Each knew that the other stood in the way of reaching that goal.

Cope took the early lead. Born outside Philadelphia in 1840, he was put on the fast track by his father, a devout Quaker who had retired comfortably from an inherited business. At 6, young Edwin wrote his grandmother an enthusiastic letter about a visit to a museum where "I saw a Mammoth and Hydrarchas, does thee know what that is? It is a great skeleton of a serpent." At 8, he was filling his sketchbook with scale drawings and precise descriptions of everything from toucans to a fossil skeleton of *Ichthyosaurus.* Although he received the highest marks for his work at school, his conduct was deemed "not quite satisfactory." Plate describes him as "a born fighter."

At age 18, Cope submitted his first paper to the Academy of Natural Sciences of Philadelphia. Naturally, the older scholars received it skeptically. But on reading, "On the Primary Divisions of the Salamandrae with Descriptions of a New Species," these scientists were so impressed that they published it in the spring 1859 *Proceedings* of the academy.

Marsh was not nearly so precocious. Born in 1831, in the northwest corner of New York State, he grew up in modest circumstances. His father had turned to farming after his successful shoe manufacturing business toppled in the depression of 1837. Marsh did demonstrate an early interest in rocks and fossils, often forgoing his chores to go collecting with a neighbor. But at age 20, he was contemplating a career as a carpenter, not a scientist.

Before forsaking his collecting hammer for a carpenter's hammer, Marsh inherited $1200. His relatives, concerned about his future, pushed him to use the money to attend Phillips Academy at Andover,

Massachusetts. Marsh wasn't terribly interested in going to school, but he was anxious to endear himself to his uncle George Peabody, who had made a fortune in business and now lived in London. Peabody was generous to relatives who showed pluck.

To demonstrate his pluck, Marsh enrolled at Andover as a 20-year-old freshman, some seven years older than most of his classmates. He drifted through his first year. Then, impelled to reconsider his life after his 22-year-old sister died in childbirth, he dedicated himself to his studies.

After graduating from Andover as class valedictorian, Marsh entered Yale with his uncle's financial support. While there, he spent much of his free time on fossil-hunting field trips. Graduating Phi Beta Kappa in 1860, he entered Yale's Sheffield Scientific School for graduate study. The following year, at age 30, he published his first paper, a look at the gold fields of Nova Scotia. His second paper made a splash. After Louis Agassiz, the world's leading authority on fossil fishes, misidentified a specimen, Marsh wrote an article correctly refuting him.

By the time Marsh refuted Cope's restoration eight years later, the two men had published many more papers and distinguished themselves in the scientific community. In 1860, the 20-year-old Cope studied comparative anatomy with Leidy at the University of Pennsylvania and recatalogued the Academy of Natural Sciences' herpetological collection. Over the next two years, he wrote 22 papers. In 1863, his Quaker father, hoping to keep him out of the Civil War, sent him to Europe to continue his education.

In Germany, Cope met Marsh, who was studying at the University of Berlin. Marsh later wrote: "During the next five years, I saw him often and retained friendly relations with him, although at times his eccentricities of conduct, to use no stronger term, were hard to bear. These I forgave until the number was approaching nearly the Biblical limit of seventy times seven."

Cope's conduct was indeed eccentric. He suffered severe bouts of melancholy. He also engaged in whirlwind spurts of activity bordering on the manic. After returning to America in 1864, he wrote numerous scientific papers, including 26 in 1868 alone. By this time he had married, had a child, taught at Haverford College, resigned his professorship, hunted fossils in Maryland and North Carolina, and moved to Haddonfield, New Jersey.

Haddonfield, home of *Hadrosaurus foulkei*, was rich in fossil beds. Cope found "seven huge Saurians" there in a short period of time, including an 18-foot leaping dinosaur that he called *Laelaps aquilungus*. When Marsh learned of his friend's discovery in the spring of 1868, he went to Haddonfield. The two men spent a pleasant week together looking for mosasaurs, plesiosaurs, and dinosaurs.

By this time Marsh had become America's first professor of paleontology, thanks largely to George Peabody. Marsh had convinced his rich uncle to endow a museum of natural history at Yale. There was no formal quid pro quo. Marsh had strong credentials for the post. But full professorships were usually given to older, more experienced scholars.

Shortly after hunting fossils with Cope in Haddonfield, Professor Marsh boarded the new Union Pacific Railroad and made his first fossil-hunting trip in the West. In Antelope Station, Nebraska, he found the bones of an extinct three-toed horse that had stood only 3 feet high. "Research proved him to be a veritable missing link in the genealogy of the modern horse," Marsh later wrote. For the first time, the academic community took serious notice of him.

The general community took notice the following year, when Marsh exposed a hoax. In 1868, an agnostic farmer from Binghamton, New York, had become fed up with Bible-thumping preachers who claimed there had been ancient giants. The farmer secretly hired a stone carver to create such a giant out of stone. He then added touches to make it look authentic and let it weather for a year before burying it near Syracuse. In October of 1869, he arranged for a friend to dig a well at the burial spot and find the phony giant.

Once unearthed, the "Cardiff Giant" became a national sensation. Fundamentalist preachers hailed the "fossilized human being" as proof that the scriptures were true. Esteemed scientists did not believe it to be a petrified human, but they nevertheless hailed it as the most remarkable archaeological artifact ever found in the United States.

The creator sold a portion of his interest in the Giant for $30,000. It was moved to Syracuse, where upward of 3000 people a day paid to see it. Unable to buy the original, P. T. Barnum commissioned a copy and displayed it in a New York City museum, where it drew huge crowds.

Enter Marsh, the longtime specialist in finding and exposing mistakes. He immediately saw that the statue was made of gypsum so soluble that it could have survived only a few years underground without dissolving. "The whole thing is a fraud," he pronounced, pleasing the perpetrator of the hoax but embarrassing the scientists and clergy who had been fooled.

While Marsh was making a name for himself, Cope was champing at the bit to get out West. Unlike Marsh he had neither an uncle nor a university to fund the trip. Cope did, however, have a farm that his father had bought years earlier. Cope the elder agreed to sell the property to pay for a western expedition.

Cope didn't make the expedition until the fall of 1871. Once past the Mississippi, he spent much of his time in the Kansas chalk fields. There he joined forces with Charles H. Sternberg, a young fossil collector who would soon make his own mark in the world of paleontology. Despite frequent gales and sandstorms, the duo unearthed the largest sea turtle found to that date, an 800-pound fish that Cope named *Portheus* ("bulldog tarpon"), and a new mosasaur, *Tylosaurus,* that had jaws big enough to swallow *Portheus.* "I secured a large proportion of the extinct vertebrate species of Kansas, although Prof. Marsh had been there previously," Cope crowed.

Marsh had been through much of the West previously. In 1870, the professor and a group of 12 students from Yale spent almost six months in Nebraska and Wyoming. They started in the summer at Fort McPherson in Nebraska, where they met Buffalo Bill Cody. Marsh and Cody became fast friends, and whenever Buffalo Bill took his show east, he visited Marsh's classes in New Haven.

Escorted by the Fifth Cavalry, the Marsh party mounted Indian ponies taken in battle from the Cheyenne and headed for Wyoming. Riding 14 hours a day, they faced hailstorms, swarms of mosquitoes, and prairie fires that the Sioux set to deter them. From Fort Bridger in southwest Wyoming, they set out for a legendary valley of huge petrified bones. They didn't find it, but they did find a bone yard rich with mammal remains.

Before heading back in December, Marsh shot a buffalo, discovered a mosasaur burial ground, and found a most unusual hollow bone. "The bird characters were there, but such a joint no known bird possessed, as it indicated a freedom of motion in one direction that no well-constructed bird could use on land or water," Marsh wrote. He concluded that the bone had belonged to a giant pterodactyl, the flying reptile of Mesozoic times.

The following year the professor discovered remains of the first birds to be recognized as having teeth. This led Thomas Huxley to write: "The discovery of the notched birds by Marsh completed the series of transitional forms between birds and reptiles and removed Mr. Darwin's proposition 'that many animal forms have been utterly lost' from the region of hypothesis to that of demonstrable fact."

Relations between Marsh and Cope had grown strained in the years following Cope's mistaken reconstruction of the *Elasmosaurus*. Each took potshots at the other whenever possible. But it wasn't until 1872 that war broke out between the two.

Cope, who had attached himself to a U.S. Department of the Interior geological surveying party, triggered the hostilities by daring to invade Wyoming's Bridger Basin, which Marsh had been draining of fossils for two years. Problems immediately arose. Each side spied on the other. Marsh bribed telegraphers for news of Cope's finds. One of Cope's scouts regularly reported new discoveries to Marsh. And Cope paid Marsh's men to lead him to their leader's sites.

Trickery was also common. After observing Marsh's party leave a site empty-handed, Cope decided to try his luck there. He found a skull and loose teeth, leading him to describe a new species. Some time later, when he saw an identical skull with very different teeth, he realized he'd been fooled. Marsh's men had intentionally "salted" the site with the bones of two different species to mislead him into the embarrassing misidentification.

Nathan Reingold, author of *Science in Nineteenth-Century America,* describes Cope and Marsh as "robber barons trying to corner the old-bones market." During their first years out west, the pair primarily found the old bones of mammals, fish, and lizards. Eventually, the bones and the stakes grew considerably larger.

In 1877, Arthur Lakes, a teacher and part-time fossil collector, found some giant bones embedded in rock near the town of Morrison, Colorado. Western collectors knew what to do when they discovered a promising site: contact Marsh or Cope, each of whom was willing to pay for good fossils or good leads to fossils. Lakes sent Marsh a letter with sketches.

While waiting for a response, Lakes found many more bones. He sent boxes of these fossils to Marsh at Yale and Cope in Philadelphia for identification. Marsh responded first. He sent Lake a check for $100 and told him to keep the location of the find a secret. There were dinosaurs in them thar hills. Marsh then published a paper on a new dinosaur, *Titanosaurus montanus,* "giant mountain reptile." (After learning that this name had already been used to describe a dinosaur found in India, Marsh changed the name to *Atlantosaurus,* "Atlas reptile.")

When he learned that Lakes had also written Cope, Marsh sent his chief field collector to Colorado to make a deal with the teacher. Lakes subsequently wrote Cope and asked him to forward to Marsh the bones Lakes had previously sent to Philadelphia. Cope was angry at being bested by his rival, but he acquiesced. Marsh named these bones *Apatosaurus ajax* ("deceptive reptile").

Soon the tables were turned. Cope received specimens from another teacher/collector in Colorado, O. W. Lucas. Cope identified them as the bones of a different dinosaur, an herbivore far more spectacular than that found by Lakes. He named it *Camarasaurus* ("chambered reptile").

Cope promptly hired Lucas, who found other important remains for him. Hearing of this, Marsh sent his own people to the site to look for dinosaurs and woo Lucas to their camp. They failed to recruit Lucas but did find several excellent dinosaur skeletons at nearby Felch Quarry.

In July 1877, Marsh received the following letter from Laramie, Wyoming:

> I write to announce you of the discovery . . . of a large number
> of fossils, supposed to be those of Megatherium, although there
> is no one here sufficient of a geologist to state for a certainty.
>   . . . We are desirous of disposing of what fossils we have, and
> also the secret of the others. We are working men and not able to
> present them as a gift, and if we can sell the secret of the fossil bed
> and secure work in excavating others we would like to do so.

After a description of the bones, the letter concluded: "We would be pleased to hear from you, as you are well known as an enthusiastic geologist and a man of means, both of which we are enthusiastic of finding—more especially the latter."

The signers of the letter, who identified themselves as "Harlow and Edwards," sent a few fossil samples in good faith. Realizing that these were dinosaur bones, Marsh sent the westerners a check for $75 and asked them to keep digging. In a second letter, "Harlow and Edwards," who turned out to be railroad men named Reed and Carlin, noted that there were others in the area looking for bones. Worried about losing out on the find, Marsh dispatched an emissary, Samuel Williston, to check out the site. Williston reported that this site, known as Como Bluff, was far superior to the Colorado sites discovered by Lakes and Lucas. The bones, he wrote, "extend for seven miles & are by the ton."

Marsh quickly put Reed and Carlin on his payroll and sent them a contract under which they agreed to "take all reasonable precautions to keep all other collectors not authorized by Prof. Marsh out of the region." In the first year, Marsh's forces excavated approximately 30 tons of dinosaur bones. The professor quickly wrote a paper describing and naming for the first time such dinosaurs as the plated *Stegosaurus* ("roof reptile"), the carnivorous *Allosaurus,* and *Nanosaurus* ("dwarf reptile"), which was only 2 to 4 feet long.

Cope, of course, wanted to enter this fertile region. He sent spies to see what was going on and then traveled there himself. At about this time, Carlin and Reed had a disagreement and parted company. Carlin opened up his own quarry and sent his finds to Cope.

Over the next ten years, the Wyoming fossil rush was as exciting if not quite as lucrative as the California gold rush had been 30 years earlier. Marsh and Cope continued to mine Como Bluff and other fossil fields in the region for dinosaurs. They apparently thought nothing of spying, bribing, intimidating, and stealing each other's employees. When Marsh's man Reed finished digging at a site, he destroyed all remaining bones so that Carlin could not come in and ship them to Cope.

Who won? Marsh's collectors uncovered 26 new species of dinosaurs at Como Bluff alone. Cope found only a handful of species at the Wyoming site, but he was more successful than Marsh in Colorado. For the record, over their lifetimes Marsh described 75 new dinosaur species to Cope's 55. In his comprehensive chronicle of the Cope/Marsh feud, *The Bonehunters' Revenge,* David Rains Wallace notes that 19 of Marsh's descriptions and 9 of Cope's descriptions are considered valid today. Cope described fewer dinosaur genera than Marsh, but he described 1282 genera or species of North American fossil vertebrates, while his rival described 536.

Science was the true winner. The Cope and Marsh forces introduced more than new species. They introduced new methods of paleontological exploration and excavation. Despite the inexcusable destruction of fossils, the western collectors were also responsible for developing techniques for preserving and shipping bones that are still used today. The papers Cope and Marsh published formed the backbone of the study of vertebrate paleontology as the century turned. The skeletons their men found remain on display in many museums around the world. Reed, for example, unearthed a nearly complete *Apatosaurus*—the correct generic name for what most people know as the brontosaurus. It remains a major exhibit at Yale's Peabody Museum.

Sadly, success in the field did not bring an end to Cope vs. Marsh. In the 1880s, Cope lost most of his money by investing foolishly in silver mines. Short of funds, he was unable to collect fossils or finish many publications. Meanwhile, Marsh, over Cope's objection, was elected president of the National Academy of Sciences. The professor was also named vertebrate paleontologist to the U.S. Geological Survey, a position that assured him of funds to continue hunting for bones.

Already reeling from his own misfortune and Marsh's good fortune, Cope suffered a knockout punch when the Department of the Interior demanded that he turn over everything he had collected while surveying for the government to the U.S. National Museum. Cope, who had been an unpaid volunteer on many surveying expeditions, had assumed that his finds belonged to him. He sensed that Marsh had orchestrated this unwarranted action.

On his way down, Cope took one last swing at his longtime nemesis. Williston and others employed by Marsh had long complained that the professor had taken credit for much of their work. Cope fed this information to a friend at the *New York Herald*. In 1890, the paper ran a lengthy story damning Marsh. The bad publicity eventually led to a reduction of funding for the Geological Survey and Marsh's resignation from his post there. As expected, Marsh struck back. The *Herald* published his rebuttal to Cope's charges, in which he recounted the 21-year-old story of the *Elasmosaurus* and charged that Cope had once broken open crates of fossils that belonged to him.

Broken financially, Cope supported himself by taking a professorship at the University of Pennsylvania and selling many of his remaining fossils to the American Museum of Natural History in 1895 for $32,000. He died in 1897 at the age of 57.

Most of Marsh's finds remain at the Peabody Museum. The specimens he collected with U.S. Geological Survey monies went to the Smithsonian. He died in 1899 at the age of 67.

Each man did receive an important honor shortly before his death. In 1895, Cope was elected president of the American Association for the Advancement of Science. In 1897, Marsh received the Cuvier Prize, the highest award in his field. The French Academy presented this honor every three years. Marsh was only the third American recipient, following Louis Agassiz and Joseph Leidy.

Edward Drinker Cope is one of Peter Larson's heroes. Unlike Marsh, who was affiliated with and funded by a university and the government, Cope was an independent contractor who bankrolled his own operations. Although he might be defined today as a commercial collector, he was first and foremost a brilliant scholar. "He was a little guy fighting the greater powers, the underdog," Larson says.

Larson notes that he and Cope had something else in common besides funding their own projects and working without institutional attachment. In 1889, the federal government had demanded many of the prize fossils Cope had collected over the years. In 1992, history seemed to be repeating itself with the government's seizure of Sue.

By the time the Eighth Circuit heard oral arguments in St. Paul on October 14—five months to the day after the seizure—Larson clearly felt he was the underdog. The institute had run up over $100,000 in legal fees trying to get Sue back. Business had also suffered. Many old clients and prospective clients were keeping their distance, waiting to see how the courts decided the matter.

On November 2, the Eighth Circuit announced its decision. Referring to Judge Battey's three-day hearing to determine temporary custody of Sue, Judge Magill wrote: "Not unlike the dinosaur in size, this hearing built up 628 pages of transcript, with 14 witnesses, and 114 exhibits." He then noted that existing case law required the Eighth Circuit to give Judge Battey's analysis of the testimony the benefit of the doubt, unless he made some obvious mistake.

Were Judge Battey's findings clearly erroneous? No, said Magill. Therefore: "We affirm the district court's order naming the South Dakota School of Mines and Technology as custodian of the fossil . . . and remand for further proceedings on the merits." Sue would stay where she was until Judge Battey ruled on the institute's lawsuit asking for a determination of who had the superior possessory interest in the bones, the institute or the government.

Larson was again devastated. Sue would be spending the winter inside a storage container inside a machine shop inside a garage—like some precious Fabergé egg within another egg. After 21 months of touching Sue and talking to her, Larson had put flesh on her bones, brought her back to life. She was no longer a fossil to him. She was a living being sharing the stories of her past, the most intimate details of her existence. Their relationship transcended the physical; it was spiritual as well. Now, both her body and her soul were in jeopardy. And all he could do was stand outside the garage and tell her that everything would be all right.

Duffy reacted to the decision in lawyerly fashion, referring to Sue as "it" instead of by her given name. "[The] ruling affects only the temporary custody issue," he told the press. "I am disappointed that for the winter it will be kept in an unheated garage, but the bulk of the case remains before us." He added that he had confidence in the jurist who had to date ruled against the institute at virtually every turn. "The case will require a lot of study and hard work, and Judge Battey is very good with difficult legal issues," he said.

There was no need for Schieffer to spin or flatter. The Eighth Circuit had given him everything he wanted. "Obviously we were pleased with the decision," he said. "I would like to claim it was because of some brilliant lawyering, but the facts were pretty straightforward."

The Cheyenne River Sioux tribe also expressed satisfaction with the court's decision. "This will be helpful to us," Steve Emery, the tribal attorney general, told *Indian Country Today*. He added that the tribe planned to file an action for Sue in tribal court in the near future. "If the tribe is able to vindicate all its interests, the fossil would likely become the focal point of a tribal museum." Pat Leiggi had agreed to help train the tribe in the curation of fossils.

On the day following the decision, U.S. voters elected William Jefferson Clinton the nation's forty-second president. By chance, that same day, Casey Carmody, the Seattle mineral and fossil dealer who was friendly with Hendrickson, received a letter from Clinton. Weeks earlier, Carmody had been invited to participate in a televised "town meeting" with Clinton, the candidate. She had told him about Sue on the air and said she felt the government had abused its seizure laws. After the meeting, Clinton had made a point of talking to her. She had then sent him material about the case.

In his letter to Carmody, Clinton wrote: "The information you sent me regarding the seizure of fossils and records pertaining to 'Sue' seems thorough and insightful. Due to the increasing demands of my schedule, I've asked my staff to review it."

Neither Carmody nor the institute interpreted this letter as an endorsement of their position, but they did see positive ramifications in his ascension to the presidency. Clinton, like every president who had preceded him, would soon install people of his own liking in many appointive positions, such as U.S. attorney. Schieffer, a Republican appointee, was already on shaky ground. The Senate had never confirmed him; he was still the acting U.S. attorney. He was a prime candidate to be replaced by the new Democratic administration.

No one could predict how a new U.S. attorney would handle this case on inheriting it. But the institute's chances of reaching a "common understanding" with the government would undoubtedly be better with a different prosecutor. Some observers felt that Schieffer had backed himself into a corner with his dramatic raid and subsequent public statements and that the only way he could save face and prove the editorial writers and the protesters wrong was by winning permanent custody of Sue and indicting the Larsons, not by settling the matter.

Schieffer offers a different explanation: that the law supported his position that the institute had no claim to Sue. "It was a no-brainer," he says. As for the criminal case, it would be up to the grand jury to determine whether indictments were warranted.

With the clock running out on his tenure, Schieffer wanted nothing more than to end the matter quickly. "We can hopefully proceed in a straightforward manner and let it be resolved by the legal process, where it rightfully belongs," he said following the ruling.

In an effort to effect a swift resolution on the institute's original lawsuit to determine who had the superior possessory claim, Schieffer had filed a motion for summary judgment on October 28. Such a motion is based on a party's assertion that there is no genuine issue as to any material (or, essential) fact in the case. Therefore, there is no need for a trial before a jury to elicit any facts. Instead, the judge can render a final decision based solely on the applicable case law and statutes.

What were these indisputable material facts in *Black Hills Institute (Plaintiff) v. The United States of America, Department of Justice (Defendant)*? According to the government:

1. The subject of plaintiffs' claim is a fossilized skeleton of a *Tyrannosaurus rex* approximately 65 million years old.

2. The land from which the fossil was taken is Indian trust land within the exterior boundaries of the Cheyenne River Sioux Reservation. . . .

3. Title to the above land is held by the United States.

4. In 1969, Maurice A. Williams received a beneficial interest in the above land by virtue of a document which on its face states "That the UNITED STATES OF AMERICA . . . hereby declares that it does and will hold the land above described (subject to all statutory provisions and restrictions). . . . " [*Translation:* the government was holding the land in trust for Williams.]

5. The fossil at issue in this case was removed by plaintiffs without the knowledge or consent of the United States.

In a brief supporting this motion, Schieffer presented the cases and statutes that, he argued, warranted summary judgment in the government's favor.

Confronted with a motion for summary judgment, an opposing party has two choices. It can argue that there are material facts in dispute and that therefore the motion must be dismissed and the trial should go forward. Or, it can make its own motion for summary judgment—agreeing that there are no facts in dispute, but arguing that the case law and statutes support a decision in *its* favor. The institute chose this second alternative.

After disagreeing on virtually everything for six months, the institute and the government were now saying that they agreed on the underlying facts of the case. Judge Battey would not have to hold a trial or hearing to determine the facts. He could simply read each party's brief and listen to each one's argument explaining why it was entitled to victory based on the cases and statutes it cited.

Or could he? "Rancher Stakes His Claim to Sue" said the headline in the December 2 *Rapid City Journal*. "Just when you thought you understood the case of the *Tyrannosaurus rex* named Sue, along comes Maurice Williams, who now says the famous fossil belongs to him," began reporter Harlan's story. Williams had remained on the sidelines while the government, the Sioux, and the institute had wrangled over custody of the dinosaur found on his property almost two and a half years earlier. Now, one day after the institute had filed its own motion for summary judgment, he attempted to enter the case. He went to Judge Battey's court seeking leave to file a "friend of the court" brief.

The gist of the brief was simple: Williams claimed that he—not the government, not the Cheyenne River Sioux tribe, not the institute—was "the current, legal owner of the *Tyrannosaurus rex* named Sue." In the brief he asserted that the institute had "hoodwinked" him out of a fossil "worth millions of dollars." He explained that he had accepted the $5000 check from the institute "as payment for egress to and waste committed on his trust land in the excavation of Sue," not for the fossil itself. In an apparent contradiction to his comments on the videotape, he claimed that he allowed the institute to remove Sue only to clean and prepare her for a later sale. In a later interview, Williams reaffirmed these arguments. "I thought [the institute] knew at the time a check isn't a contract," he said.

Williams further asserted that the trustee of his land, the U.S. government, had "breached its fiduciary duty" to him by claiming the fossil for itself instead of protecting his interest in it. The Antiquities Act of

1906 did not allow the federal government to take possessory interest in fossils found on Indian lands, he argued.

What did Williams want from Judge Battey? In his brief he argued that ownership of the fossil could not be determined without a full trial and his (Williams's) participation. Therefore, he said, the judge must deny both the institute's and the government's motions for summary judgment.

Williams's eleventh-hour ownership claim inspired Don Gerken, a columnist for the *Hill City Prevailer*. Tongue firmly in cheek, he invited his readers to join the fray in a "Stake Your Claim to Sue!" contest. "You too can file your claim for ownership of Sue," wrote the columnist. "It's fun!!! It's easy!!!" He then asked readers to state why they deserved the dinosaur. Among the choices he offered were: "I once saw the fossil," "I'm related to Sue," "I voted Republican in the last election," "I'm a minority," "I'm not a minority," and "I need the bucks."

A few weeks later, those actually claiming Sue played: "I think Sue is mine because . . ." on national television. "It may be the custody battle of the century," intoned anchor Sam Donaldson of ABC News's *Primetime Live*. "Scientists, Indians, the U.S. government—they're all trying to lay claim to a bunch of old bones. Well, not just any old bones . . . ."

On camera, correspondent Sylvia Chase asked the same question of four grown men.

Chase:	Who owns Sue?
Maurice Williams:	If the laws of the land mean anything, I own it.
Gregg Bourland:	We believe we do, the Cheyenne River Sioux tribe.
Kevin Schieffer:	The public owns Sue.
Peter Larson:	The Black Hills Museum of Natural History Foundation.

These quadraphonic "I do's" brought to mind four children squabbling over the last piece of pie, each child intractable.

Chase asked Williams why he claimed ownership. "Isn't it the money?"

"It's always the money," said Williams.

It wasn't the money that had motivated the government to seize Sue. Schieffer had secured that initial search warrant in May, in large part by persuading Judge Battey that there was probable cause that the institute had taken the fossil in violation of the Antiquities Act. For eight months he had continued to invoke the act when explaining to the public why he seized the dinosaur. He had also presented the institute's violation of the act as the primary of two theories in support of his motion for summary judgment in the case for ownership before Judge Battey.

Was the Antiquities Act applicable to fossils?

In 1974, Dr. Farish Jenkins, a respected Harvard paleontologist, was arrested and charged with a federal crime under the act because he "did appropriate, excavate, injure, or destroy a historic or prehistoric relic or monument, or an object of antiquity situated on lands owned or controlled by the Government of the United States." The "objects of antiquity" in this instance were fossils.

Jenkins replied that he hadn't known he was on federal land during a collecting expedition in Montana. He had, he said, inadvertently crossed an unmarked boundary onto Bureau of Land Management property. Academic and commercial collectors alike could sympathize. The boundaries between private and public lands were often unmarked, and maps were often outdated or difficult to read.

But Jenkins went beyond claiming ignorance. He challenged the constitutionality of the Antiquities Act. In doing so, he cited a previous case involving Native American masks in which the Ninth Circuit Court of Appeals had held the act "fatally vague" after the defendants had raised the constitutionality question. The district court judge in Montana agreed with Jenkins and dismissed the charges.

Until the seizure of Sue 16 years later, *United States of America v. Farish Jenkins* was the only reported case in the 86-year history of the act involving fossils rather than human artifacts such as Indian relics and, occasionally, shipwrecks on submerged federal land. In the absence of previous cases in which the act had been applied to fossils, how could Schieffer go before a federal judge and request a search warrant based on the institute's violations of the act? And how could he argue that the institute had no right to Sue because the act did not recognize private ownership of fossils excavated and removed from lands owned or controlled by the United States?

In fairness to Schieffer, the request for the search warrant cited violations of two other federal statutes in addition to the Antiquities Act. Still, in his public statements and in his brief supporting his motion for summary judgment in the case before Judge Battey, the U.S. attorney appeared to rely principally on the act. In the brief, Schieffer cited a 1956 Department of the Interior memorandum. The memorandum acknowledged that the Antiquities Act didn't specifically refer to fossils but reasoned that the government's long-term practice of granting permits for fossil collecting created "a strong presumption of the validity of the practice." As a result, the memorandum concluded that the official position of the Department of the Interior was that fossils were covered by the act.

If the Interior Department had remained silent on this question after 1956, Schieffer would have had a point. But the department had been loud and clear in recent years, said Duffy in his brief supporting the institute's own motion for summary judgment. Duffy cited Department of the Interior memoranda from 1977 and 1986 that reversed the 1956 opinion. The 1986 memorandum had stated that "analysis in these terms would also lead to the conclusion that paleontological fossil specimens were not to be included under the sections of the act establishing permit procedures and penalty provisions." Schieffer had not mentioned this or the 1977 memorandum.

Shortly after Duffy filed this brief and Williams filed his motion to enter the case as a friend of the court, Schieffer dropped a bombshell. "The United States will no longer rely on the Antiquities Act," he told the court in a brief responding to Williams's motion. Duffy was furious. "*T. rex* Fossil Finders Say Government Lied" ran the front-page headline in the *Journal.*

Duffy told the paper that the government had known for 15 years— since its own 1977 memorandum—that the Antiquities Act was designed to prevent the collection of artifacts, not fossils, from federal property. The U.S. attorney had obtained the warrant to seize Sue under false pretenses, said Duffy. "The basis for the seizure, or for continued holding of the dinosaur, no longer exists." Sue, Duffy said, should be returned to the institute.

6

IS A DINOSAUR "LAND"?

"Three million four hundred thousand."

A faint smile crossed Maurice Williams's weathered face. This *was* all about money. He had been uncertain how much the *T. rex* would command. Now he had a good sense of her worth. "It's kind of like trading horses, isn't it?" the rancher would say later. "The true value is what you can get out of it."

Maurice Williams was not naive. He was 66 years old when Sue was found. He held a college degree in animal sciences. He had once worked for the Bureau of Indian Affairs. According to unofficial reports, he owned more than 35,000 acres of land. Those who knew him said he was a tough businessman.

Williams had waited until the institute, the government, and his own Cheyenne River Sioux tribe were into the home stretch before making his move. His motion to enter the case as a friend of the court, not Duffy's citation of the two Department of the Interior memoranda, caused Schieffer to alter his strategy. "To [rely on the act] could conceivably prejudice Williams and require that he be allowed to intervene as an indispensable party," Schieffer wrote. Such intervention would raise numerous factual questions and necessitate a lengthy trial rather than a ruling based on the motions for summary judgment, the U.S. attorney knew. And such a trial might result in the determination that Williams, not the government, owned Sue.

In dropping the Antiquities Act argument, Schieffer was not dropping his claim that "the fossil is property of the United States. Period." In his motion for summary judgment, he had offered an "alternative" argument: Sue was land—real property—he asserted. Thus,

> the basic principles of real property law and Indian trust law must be applied. Under this alternative approach . . . Maurice Williams . . . had no authority to unilaterally sell or convey the trust land or any objects embedded in it without the knowledge and consent of the U.S.

"Absolutely preposterous," Williams told the *Indian Country Today*. "If the government's position was followed to its logical conclusion, the government would also own all the timber, grass, oil, gravel, coal, and other minerals on trust land." The government had never previously claimed such rights. Indeed, Williams had sold oil rights on his property to Amoco.

Williams had equally harsh words for the Cheyenne River Sioux tribe, which was still asserting ownership and threatening to bring action in tribal court. "I don't know what this world is coming to," said the rancher. "The tribe is supposed to be protecting me. It has many, many fossils of its own on its own lands. Yet the tribal attorneys are exerting an extraordinary amount of time and energy trying to steal my fossil."

Williams's enemies list didn't stop with the tribe. He went on to criticize the United States Bureau of Indian Affairs (BIA). He said he had repeatedly asked the BIA to provide him with an attorney to represent his interests in the case. "Why can't the responsible federal officials respond properly to my repeated requests for justice and direction out here in Indian country?" he asked. "The government would have a much stronger case against the Larsons if its officials would do the right thing by acting as my trustee and protecting my property interests."

This criticism called to mind an episode involving the BIA, the Sioux, and Othniel C. Marsh over 100 years earlier. In 1874, Marsh was busy with the construction of the Peabody Museum at Yale, but he could not resist the invitation of U.S. Army officers to hunt for fossils in the Badlands near the Black Hills. He arrived out West during a time of great tension. The Oglala Sioux felt that the federal government had violated an 1868 treaty.

This treaty had ended a war between the U.S. Army and the tribe, led by its fearless chief Red Cloud. It gave the Sioux, as a "permanent reservation," all of what is now South Dakota west of the Missouri River. Under the agreement, no white man could settle on the land or even pass through without the permission of the Sioux. In return, Red Cloud swore that he would never again wage war against the whites.

Red Cloud had kept his promise, despite pressure from the more militant members of his tribe, Sitting Bull and Crazy Horse. The United States, however, had reneged. In 1871, the government announced that the Northern Pacific Railroad would run through the land in question. Then, in the summer of 1874, a few months before Marsh arrived, General George Armstrong Custer led 1000 men to the Black Hills. They weren't looking for Indians. "Veins of gold-bearing quartz crop out on almost every hillside," Custer reported.

The Sioux knew that a gold rush would bring in settlers. Feeling betrayed once again, they refused to cooperate with the U.S. Indian agent at the reservation and threatened trouble. The corrupt agent exacerbated the hostilities by cheating the tribe of promised provisions. He dispensed rotten meat and threadbare blankets

In *The Dinosaur Hunters,* Plate writes that despite the warnings of frontiersmen to stay away from the reservation, "Marsh stubbornly proceeded . . . and set up camp in the midst of 12,000 surly Indians." Once established, he sought permission to enter the Badlands to search for bones. Red Cloud was suspicious. Marsh, he reasoned (inaccurately), was interested in finding gold, not fossils. When the chief balked, Marsh proposed a deal: If Red Cloud would help him get fossils, he would take Red Cloud's complaints about the agent and other fraudulent government practices back to Washington, D.C. Red Cloud agreed.

The chief assigned Sitting Bull to take Marsh and his men into the Badlands. After a snowfall delayed the journey, the Sioux had a change of heart. They had received word that two angry tribes, the Miniconjous and Hunkpapas, lay in wait for the expedition. Marsh's friends in the army also cautioned against the trip.

Marsh wouldn't quit. Hoping to persuade the tribe to reconsider its decision, he held a grand feast. After the meal, Red Cloud relented. He assigned his nephew Sword to escort the party. But again fearing the Miniconjous, the Sioux refused to go.

Undaunted, Marsh and his men went gently into the night while the Sioux slept. The unescorted party reached the Badlands safely. There they found many excellent fossils.

Red Cloud did order his men to stand guard on the buttes above the fossil hunters. When the chief learned that the Miniconjous were planning to attack, he sent word that Marsh should leave immediately. Marsh refused. A hasty exit would necessitate abandoning his haul of fossils. The professor insisted on spending one more day at the site to pack the bones safely for the trip back east. He and his men finally left only hours before a Miniconjous war party reached the spot where they had camped.

Marsh kept his promise to Red Cloud and went to Washington when he returned home. By this time the Sioux were facing starvation. The government rations were both insufficient and inedible.

The corrupt administration of President Ulysses S. Grant showed little interest in Marsh's report. The professor persisted. He pressed his case in the media, revealing ten major government frauds, including kickbacks to contractors. The administration and those businessmen benefiting from the fraud responded by personally attacking Marsh. He was, they said, seeking fame, or he had been bribed, or he was just plain crazy.

Remarkably, after several months, Marsh prevailed. The Secretary of the Interior, the head of the Bureau of Indian Affairs, and the agent on the reservation all resigned or were fired. Plate reports that Marsh's triumph made Red Cloud his friend for life. The Sioux chief sent Marsh a peace pipe and wrote: "I thought he would do like all white men, and forget me when he went away. But he did not. He told the Great Father everything just as he promised he would, and I think he is the best white man I ever saw."

Now, a century and a quarter later, Williams was willing to put his fate in the U.S. attorney's hands. After finally getting his audience, via telephone, with the Interior Department, Williams decided not to intervene in the case because, he said, "my interests would be adequately protected in the lawsuit by the U.S. government." Schieffer, however, never publicly stated that he was representing Williams's interests.

The Larsons knew exactly where they stood with the U.S. attorney. By the middle of January, the two brothers, partner Bob Farrar, chief preparator Terry Wentz, administrative assistant Marion Zenker, curator David Burnham, and three other institute workers had been subpoenaed to appear before the grand jury.

On January 19, the "Institute 9" traveled to Rapid City. The Larsons, Farrar, and Wentz—the four from the institute assumed to be under investigation—were asked only for handwriting samples, not personal testimony. Afterward Duffy complained that all those subpoenaed had been harassed, intimidated, and treated shoddily. "It's like Bosnian jurisprudence," he said. Some had to sit on the floor of the waiting room, which had only three chairs, he said. All, even those who testified early in the morning, were forbidden to leave until 6:30 PM.

Burnham said that the assistant U.S. attorney questioning him "screamed and yelled at me." He added: "I don't even know why I was there except to be abused and harassed." Ron Banks, an attorney representing some of those testifying, concurred. "It's deplorable the way they treat witnesses," he told the *Journal*.

In a letter to the paper, Schieffer said that attorneys who truly believed that their clients' rights were being violated had instant access to the federal court to stop such abuse. None had done so. He denied the charges of harassment and demanded that the paper apologize for publishing reports of intimidation.

Before the month was over, the government once again subpoenaed institute records. Peter Larson said that the subpoena required the institute to turn over almost everything but equipment and fossils—virtually all remaining business records, field notes, videotapes, and photographs. Sixty-four general and specific items were requested, including documents related to 20 tons of fossils collected in Peru. Larson suggested that the broad demand required him to send the government even the 33-year-old picture of him and Neal in front of their "museum." He posed for the *Journal* holding the photo. The institute had 24 hours to comply with the request.

During the same week that this subpoena was served, six FBI agents arrived at the institute with a court order authorizing them to photograph more than 100 different fossil specimens. "During the agents' visit . . . local residents reportedly refused to offer [the] agents coffee or even allow them to use private bathroom facilities because of their ris-

ing level of anger at the federal government's handling of . . . the case,"
noted the local *Tribune/News*.

Following the subpoenas and the grand jury appearances, a newly
established citizens' organization in the Hill City area, the Government
Accountability Group (GAG), sent a petition to the U.S. Department of
Justice and the Senate Judiciary Committee. The petition sought immediate investigation into alleged violations of the Fourth, Fifth, Sixth, and
Fourteenth Amendments by Schieffer, Zuercher, and FBI Agent Asbury
in their "conspiracy of harassment" against the institute. Charging
Schieffer and Zuercher with malpractice and malfeasance, GAG also
sought disbarment hearings of the pair. Finally, the petition also asked
the newly inaugurated President Clinton to replace Schieffer with an
interim U.S. attorney and to issue "an executive order to cease and desist
in this merciless and vindictive persecution." A second executive order to
return the subpoenaed documents and specimens was also requested.

Travis Opdyke, a writer and former FBI employee, wrote the petition. He explained that he had been moved by "the tremendous amount
of anger" engendered in the community by the government's actions.
Opdyke was married to the institute's Marion Zenker.

Experienced criminal defense attorneys don't like to second guess
their fellow practitioners. However, many will say that during an investigation they endeavor to maintain a cordial, civil relationship with the prosecutors who have the power to indict their clients. The rancorous relations
between the U.S. attorney's office and the institute and its lawyers and supporters—no matter whose fault—had the potential to sabotage future
attempts by Duffy to, if necessary, make a deal in the criminal case with
Schieffer that might benefit the Larsons. "Pat Duffy is a smart guy," Colbath
would later say. "But if you're going to use tactics like his—making your
witness a martyr in the press—you need a whole apparatus, a public relations machine, behind you. Sometimes when you're trying to make a client
into a martyr, the government is all too willing to help."

Duffy's curriculum vitae suggested that he was indeed "a smart guy."
The son of a lawyer, he was born in 1956 and raised in Ft. Pierre, South
Dakota. Although Duffy always knew he wanted to be a lawyer, he had
been counseled by his father to get real-life experience before entering
the profession. Taking the advice to heart, he enlisted in the U.S. Navy
after high school. After receiving training in Russian, he was given the

opportunity to enroll at the U.S. Naval Academy. He did enroll, but two years later he quit, returned home, and married. He eventually received his undergraduate degree from South Dakota State University. Still not ready for law school, he worked as a stockbroker for several years.

Duffy finally entered the University of South Dakota Law School in 1983. There he served as editor-in-chief of the law review. After graduating in 1986, he went into private practice doing both civil and criminal trial work. He represented the institute for the first time soon after graduation. Larson came to him after receiving a subpoena for the documents related to the government's Wyoming fossil investigation.

Never one to mince words, Duffy had confronted Schieffer soon after the seizure. "I said, 'You asshole. You lied to me,'" Duffy recalls, referring to Schieffer's prior assurance that no raid was imminent (an assurance Schieffer says he never made). Schieffer told him that such a response was necessary to "preserve the integrity of the operation," remembers an incredulous Duffy, who adds, "That's like the stuff we heard from Vietnam: 'We destroyed the village in order to save it.'"

Duffy told Schieffer that the raid was totally unnecessary. "If Kevin had just called me and said, 'Let's litigate,' I'd have promised [that Sue would be kept safe]," he said many years later.

Duffy knew he was taking a risk in playing the case out in the press. "In many cases I don't say a thing," he says. "But [here] my media presence was absolutely necessary. When Schieffer lied to me and showed up in pancake makeup, I knew this wasn't like dealing with a real lawyer. I had to fight back. This was [Schieffer's] publicity platform for a judgeship or his run for the Senate. He was writing op-ed pieces. I had to keep the temperature of the body politic such that we could get a fair trial."

Bob Chicoine, a prominent Seattle attorney who also represented Hendrickson, questions Duffy's strategy. "Being aggressive with the other side is one thing," he says. "But once it becomes personal, that's the worst thing. You are dealing with prosecutors who have an immense amount of power. It's bad enough to piss off an assistant U.S. attorney, but to piss off the U.S. attorney himself is crazy."

In Chicoine's opinion, the most effective lawyers keep themselves out of the spotlight and certainly don't shine it in the eyes of those who may decide their client's fate. "Prosecutors and judges are human, too," says Chicoine. "They don't like to have their integrity questioned in public."

Schieffer was equally baffled by Duffy's tactics. "Maybe they thought their best chance was trying the case by press release rather than by serious civil lawsuit," he says. "But it was a bizarre strategy as far as I could figure out that didn't do too much for the client. It was different than anything I ever experienced."

Federal rules prevented Schieffer from publicly discussing the criminal investigation. Once the grand jury convened, he generally adhered to those rules, refusing to respond to the attacks of Duffy and the Larson brothers' supporters. As a result, the institute won the media battle.

Schieffer is the first to admit this. "We knew public relations was going to be a huge headache no matter how you sliced it," he says. Although blessed with what he terms a "fairly thick skin," there were times when he would have liked to tell the government's side of the story to the public. "These guys had flagrantly violated the law," he says, adding, "If you start from the beginning, they knew this was not a [mere] $5000 specimen. The skullduggery started the very first day."

As the grand jury's investigation of alleged skullduggery heated up, the case to determine who owned or at least who should possess Sue seemed to be winding down. Because the government and the institute had agreed to all the material facts of the case, Judge Battey's decision would in all likelihood depend on how he answered a question that seemed more appropriate for a philosophy classroom than a courtroom: Is a fossil personal property (in this case, an object apart from the land) or real property (part of the land itself)?

Said the government: The fossil was land or an interest in land; therefore Williams did need government permission to sell it. Said the institute: The fossil was neither land nor an interest in land; therefore Williams did not need permission.

Congress's effort to regulate Indian lands dates back to the nineteenth century. In a *South Dakota Law Review* article, "Jurassic Farce," Duffy and coauthor Lois Lofgren presented a comprehensive history of such regulation. This history says as much about how society viewed Native Americans as how it viewed their lands:

> If these wards of the nation were placed in possession of real estate, and were given capacity to sell or lease the same, or to make contracts with white men with reference thereto, they

would soon be deprived of their several holdings; and . . . instead of adopting the customs and habits of civilized life and becoming self-supporting, they would speedily waste their substance, and very likely become paupers.

So said the Eighth Circuit in the case of *Beck v. Flournoy Live-Stock and Real Estate Co.* In this 1894 decision, the court was interpreting the Congressional intent behind enactment of the General Allotment Act of 1887 (GAA). The GAA gave individual Native Americans a parcel of land (or "allotment") to "enable them to become independent farmers and ranchers." Each Native American "owned" the land for the purposes of farming, grazing, and residence, but the United States held title to the allotment in trust for 25 years. During that time, the land could not be sold, transferred, or taxed. By creating this quarter-century trust period, Congress intended the Native American "to become accustomed to his new life, to learn his rights as a citizen, and prepare himself to cope on an equal footing with any white man who might attempt to cheat him out of his newly acquired property." In *Beck,* the court noted that the act protected Indians "from the greed and superior intelligence of the white man."

The Indian Reorganization Act of 1934 stopped the allotment policy. However, the act provided an indefinite extension of the trust period mandated by the GAA. It also assured that the trust be passed on to the heirs of the landowner. Apparently, Congress still believed that Native Americans needed to be protected from their own incompetence and the white man's greed.

Over the next 14 years, the Native Americans either demonstrated that they were more competent than previously believed or the white man demonstrated that he was less greedy. The petition process was instituted in 1948, when Congress gave the Secretary of the Interior "discretion . . . upon application of the Indian owners to . . . approve conveyances with respect to lands or interests in lands held by individual Indians."

"Lands or interests in lands" was the operative phrase, argued Duffy. Sue was neither, and therefore there was no need for Williams to petition, the institute argued in its brief:

Sixty-five million years ago, when Sue was alive, there was no doubt that a *Tyrannosaurus rex* was not "land." . . . After Sue died, her bones on the surface of the earth remained personal property. While the bones were lying exposed on the earth, they were certainly "movable" and thus were personal property within the meaning of South Dakota law. Likewise, after the partially buried fossil was unearthed, there is no doubt whatsoever that the *rex* was personal property.

In its brief, the government offered its own analysis as to whether or not Sue was land:

Under South Dakota law, "land is the solid material of the earth, whatever may be the ingredients of which it is composed, whether soil, rock, or other substance." . . . The analogy to minerals and other substances which, though eons ago organic, have long since become part of the earth is here well placed. Indeed, the composition of the fossil is primarily rock and mineral.

Duffy knew that forcing Judge Battey to decide whether Sue was land or personal property was to put him between a rock and a hard place. The institute's brief, therefore, offered an additional reason for deciding that its possessory interest was superior to the government's. As a matter of "public policy," the court should not void Williams's sale of Sue to the Larsons, Duffy argued:

The bottom line is that Williams, as an Indian, should have as much right to sell fossils from his land as would a white person from her land, particularly in this day and age of economic self-determination for Indians. A decision affirming the government's seizure of the fossil, would . . . rob [Indians] of the respect they are owed as individuals who can contract with respect to their own personal property.

In an odd way, Peter Larson, or at least his lawyer, had become Othniel C. Marsh—pleading the case of the Indians in return for the rights to a fossil.

Judge Battey issued his opinion on February 3, 1993. "The case has had a somewhat convoluted and checkered past," he began. After reviewing the material facts on which the government and the institute agreed, he noted: "The ultimate issue is whether [the institute] obtained ownership to the fossil while the land from which it was excavated was held by the United States in its trust capacity."

Ownership? Wasn't the ultimate issue merely who had the superior possessory interest? No, said the judge. "A permanent possessory right to the fossil is subsumed within the context of ownership."

Judge Battey refused to accept Duffy's public policy argument that Williams could ignore the relevant statutes solely because he was a competent Native American. As a result the fate of Sue would rest on the courtroom version of the old parlor room game Animal, Vegetable, or Mineral.

Under the heading, "Was the Fossil an Interest in Land?" Judge Battey wrote: "The court has found no case authority specifically holding that a paleontological fossil such as the fossil 'Sue' embedded in the ground is an 'interest in land.'" Therefore, to address the issue of whether Sue was real property or personal property, the judge had turned to the "helpful" definitions found in the South Dakota statutes:

> **Real and personal property distinguished.** Real or immovable property consists of: (1) Land; (2) That which is affixed to land; (3) That which is incidental or appurtenant to land; (4) That which is immovable by law. Every kind of property that is not real is personal.

> **Land as solid material of earth.** Land is the solid material of earth, whatever may be the ingredients of which it composed, whether soil, rock, or other substance. [This definition had been quoted by the government in its brief.]

Citing no other authority than these definitions and offering no other analysis or explanation, the judge wrote: "The court finds that the embedded fossil was an interest in land as defined by these provisions and therefore subject to the requirements of [the federal statutes]." Having made this determination, his final decision was inevitable:

> Maurice Williams did not make application for consent to the removal of the embedded fossil. [The institute] was equally

responsible to insure that consent was obtained in compliance with federal law. Without such consent, the attempted sale of the fossil "Sue" embedded within the land is null and void. [The institute] obtained no legal right, title, or interest in the fossil as severed since the severance itself was contrary to law.

It would have been a relatively simple matter to have applied for the removal of the alienation restraint. Had there been such an application and secretarial approval, all these months of contention could have been avoided. [The institute] must assume much of the fault caused by the failure to conform their conduct to the federal laws and regulations. [It] should have investigated the status of the land involved. They ran the risk of this unlawful taking of the fossil from Indian land by not having done so.

"Judge Rules *T. rex* Not Institute's," shouted the front-page headline in the next day's *Journal.* "Judge Wrecks Plans for *T. rex,*" said the *Minneapolis Star Tribune.* "Dinosaur Fossil Belongs Not Just to the Ages but to the Government," observed *The New York Times.*

The Larsons, unaware that the judge's ruling was imminent, were on their way to a gem and mineral show in Tucson, Arizona. "We were in Truth or Consequences, New Mexico, of all places, when we got the news," Peter recalls. It was official now: After thousands of man hours of work and an estimated $209,000 in out-of-pocket expenses (not to mention more than $100,000 in attorneys' fees), they had nothing. Duffy spoke for himself and the institute. "I'm very, very disappointed," he told the press. "We will appeal immediately."

Schieffer also spoke to the media. "Notwithstanding all the media hype, this is a pretty clear-cut issue. I just don't see how the court could have ruled differently."

The judge's opinion was clear-cut on one issue: The institute did not own Sue. But Battey had not stated who did ultimately own the fossil. In addition to the government, Williams and the Cheyenne River Sioux were still claiming Sue. In his post-opinion talk with the press, Duffy opined that, unless the institute won its appeal, ownership would revert to Williams because the judge had declared the sale null and void. He added that Williams would have to pay back the $5000 the institute had paid for Sue.

Schieffer wasn't so sure that Williams owned Sue. "This was an area where the federal government had internal inconsistencies. You could have gone any one of three ways," he says. Various laws supported giving Sue to Williams, the tribe, and the federal government—which, as a sort of trustee of antiquities for the American people, would make it available for study and display. Although Schieffer had his own preference and would later lobby for it, he did not believe the decision was his to make. "The question needs to be resolved by the administration," he said after Judge Battey ruled.

The usually loquacious Williams initially declined comment. A few days later, when reached by reporter Pamela Stillman of *Indian Country Today,* he said that he owned Sue. "The laws are very clear."

The chairman of Williams's tribe disagreed. "Both the landowner [Williams] and the institute committed a crime on the reservation by digging up the fossil," Gregg Bourland told Stillman. "[Battey's opinion] goes to show that justice has been served." Steve Emery, the tribal attorney, added that the tribe would soon file an action in tribal court for ownership of Sue; the tribal council had approved such a filing six weeks earlier but, at Schieffer's request, had waited to act until Judge Battey's decision.

Before the week was over, Duffy had filed a brief requesting that Judge Battey reconsider the ruling. "This court's decision is on a collision course with Indian self-determination," wrote Duffy.

Emery also invoked Indian self-determination in his argument on behalf of the tribe. "We had gold in the Black Hills, and we lost the Black Hills. Now we've got paleontological gold and they came in and ripped that up."

One could imagine *Primetime Live*'s Sylvia Chase returning to the Black Hills after Judge Battey's opinion and asking the same question of Larson, Schieffer, Bourland, and Williams that she had asked months earlier. "Who owns Sue?" The television correspondent would have found that nothing had changed. All four still claimed ownership. The parties seemed much less concerned with the noble virtue of self-determination than they were with self-aggrandizement.

As the inharmonious quartet continued to fiddle around, the bones of contention continued to sit, if not crumble, at the School of Mines. Said the institute's Zenker: "This is probably the premiere paleontological specimen of the world today, and to have it hidden away in the dark in a giant metal container is not at all what we had in mind or supposedly what the government had in mind, but that's what's happening."

And, remarkably, that's what would continue to happen for almost five more years.

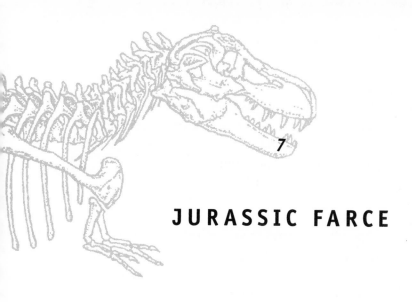

7

JURASSIC FARCE

"Four million dollars."

Fred Nuss, a tan, wiry man in his forties leaned forward in his chair. The fossil hunter hadn't come from his home in Otis, Kansas, to bid on the dinosaur. He already had a *T. rex*, although it wasn't quite as complete as Sue.

Unlike Peter Larson, Nuss had decided to sell his dinosaur. He and his business partner Alan Detrich were asking $10 million for the specimen, which they called, "Z. rex" after a rancher named Zimmerscheid. Their dream was the academic community's nightmare. The entrepreneurs didn't care if the buyer was a museum or a private collector or if Z. ended up in the United States or abroad. Nuss had once sold a mosasaur skull to actor Charlie Sheen for $30,000.

John Tayman, a writer who covered the auction for *Outside* magazine, reported: "When each million-dollar threshold was reached, Nuss and his team let out a polite whoop and then sat back giggling while the hundreds of thousands flew by."

While the institute lawyers constructed their appeal of Judge Battey's decision, Peter Larson reconstructed the life of his other *T. rex*. Each of Stan's scars told a story. "We surmise that [he] scuffled for territory, fought over food, and engaged in other behavior similar to today's carnivores," Larson wrote.

Stan's "pathologies" included several broken and healed ribs and a scar the same size as a *T. rex* tooth. He also appeared to have suffered and survived a broken neck. In the process of healing, two vertebrae fused together and a third became immobilized by extra bone growth. His cheeks also showed evidence of healed injuries.

Looking at Stan's skull, Larson the paleontologist turned phrenologist. "Most chilling is a healed injury on the back of the braincase," he wrote. "Through the back end of the skull we found a circular hole more than 1 inch in diameter—into which a *T. rex* tooth fits nicely. The hole ends at a spot where a large chunk of bone (2 inches by 5 inches) actually broke away. Amazingly, Stan lived through this incredible injury because a thin layer of bone sealed the broken surface."

Stan Sacrison, the amateur who had found this *T. rex*, had also helped collect Z. rex. After spending months on their own looking for such a dinosaur without luck in 1992, Nuss and Detrich had driven to Buffalo, South Dakota, to see if Sacrison and his twin brother Steven might help them. Stan, an electrician by day, and Steven, a construction worker and part-time grave digger, had enjoyed great success digging for fossils. The twins, in their mid-thirties at the time, introduced Nuss and Detrich to Zimmerscheid, who had found bones on his land outside of town.

Like Sue Hendrickson, the brothers seemed to have been born with the ability to find fossils. The lanky, sandy-haired Stan was just 8 years old when he found his first dinosaur bone, the vertebra of a *Triceratops*. He has been hunting for dinosaurs ever since.

In 1987, Sacrison first spotted the bones of the *T. rex* that would eventually be named after him. His description of first seeing vertebrae weathering out of a cliff calls to mind Hendrickson's discovery of Sue.

In need of an expert opinion, Sacrison called the Museum of Geology in Rapid City for help. The museum's "expert" told him he'd found a *Triceratops* and opined that it wasn't worth excavating. Sacrison took the advice. As a result, nothing was done for five years.

In 1992 Sacrison happened to tell a friend about the find. The friend suggested that he call Peter Larson. Larson looked at the bones and, again, immediately knew that they belonged to a *T. rex*, not a *Triceratops*. Thus began a mutually beneficial relationship between the Sacrisons and Larson. Stan Sacrison credits Larson with helping him hone his collecting skills and enabling him to indulge his passion for collecting by paying him for his finds.

The contemporary Larson/Sacrison relationship echoes the nineteenth-century relationship between Edward Drinker Cope and his protégé Charles H. Sternberg. Sternberg was born in 1850 in upstate New York. Like Sacrison, he developed a love of fossils at an early age. In 1867 his family moved to a farm on the Kansas prairie. There, he began collecting in earnest.

When he was 10, Sternberg fell 20 feet from a ladder, permanently injuring his left leg. He walked with a pronounced limp for the rest of his life. This injury did not prevent him from pursuing his dream and eventually earning a living excavating and selling dinosaurs and fossils to museums and individuals around the world. In his 1909 autobiography, *The Life of a Fossil Hunter*, he recalled: "I made up my mind . . . I would make it my business to collect facts from the crust of the earth; that thus men might learn more of 'the introduction and succession of life on our earth.' My father was unable to see the practical side. . . . He told me that if I had been a rich man's son, it would doubtless be an enjoyable way of passing my time, but as I should have to earn a living, I ought to turn to some other business."

The Life of a Fossil Hunter painstakingly details the trials and tribulations of the men who earned a living hunting bones in the second half of the nineteenth century. In 1876, while a student at the Kansas State Agricultural College, Sternberg attempted to join an expedition led by Benjamin F. Mudge, a geology professor who collected specimens for Othniel C. Marsh. Mudge turned him down. Sternberg later wrote: "Almost with despair, I turned for help to Professor E. D. Cope of Philadelphia. . . . I put my soul into the letter. . . . I told him of my love for science and of my earnest longing to enter the chalk of western Kansas and make a collection of its wonderful fossils, no matter what it might cost me in discomfort and danger."

Confessing that he was too poor to go at his own expense, Sternberg boldly asked Cope to send him $300 "to buy a team of ponies, a wagon, and a camp outfit, and to hire a cook and driver." Cope responded quickly. Sternberg writes: "When I opened the envelope, a draft for $300 fell at my feet. The note that accompanied it said: 'I like the style of your letter. Enclose draft. Go to work,' or words to the same effect."

Sternberg went to work. In the Kansas chalk, he toiled from dawn to dusk, "forgetting the heat and the miserable thirst and the alkali water, forgetting everything but the one great object of my life—to secure from the crumbling strata of this old ocean bed the fossil remains of the fauna of Cretaceous Times."

The constant labor, however, weakened him. "I fell a victim to malaria, and when a violent attack of shaking ague came on, I felt as if fate were indeed against me." In the middle of a shaking fit, he found a fine mosasaur, which Cope would later name *Clidastes tortor* because of its flexible backbone. "Forgetting my sickness, I shouted to the surrounding wilderness, 'Thank God! Thank God!'" Sternberg wrote.

When Cope traveled west a few months later, in August 1876, Sternberg and his assistant J. C. Isaac met him in Omaha. "I remember him watching me with astonishment as I limped along the street on my crippled leg. At last turning to Isaac, . . . he asked, 'Can Mr. Sternberg ride a horse?'"

Isaac answered that he had seen Sternberg "mount a pony bareback and cut one of his mares from a herd of wild horses." That satisfied Cope. Sternberg remembered: "When we got to Montana, he gave me the worst-tempered pony in the bunch."

A bad pony was the least of Sternberg's worries once the party reached the Badlands just a few weeks after General Custer and his men fell at Little Bighorn. Indians did not present a problem, but the terrain did. Sternberg recalled a frightening slip while crossing a sandstone ledge where a perpendicular escarpment dropped downward for 1000 feet. "God grant that I may never again feel such horror as I felt then, when the pick, upon which I had depended for safety, rebounded as if it had been polished steel, as useless in my hands as a piece of straw. I struck frantically again, but all the time I was sliding down with ever-increasing rapidity toward the edge of the abyss . . . and certain and awful death below."

Sternberg survived—"To this day I do not know how," he wrote—to face more mundane troubles: "We were tormented by myriads of black gnats, which got under our hat rims and shirt sleeves, and produced sores that gave rise to pus and thick scabs. . . . We were forced, for lack of something better, to cover our faces and arms with bacon grease."

The results of the expedition made the torment tolerable. In the Cretaceous formations of Montana, Cope and Sternberg found dinosaurs.

Judge Battey might have found Sternberg's century-old observations about the discovery interesting. "Fossil bones always partake of the characteristics of the rock in which they are entombed, and here they were quite hard when we got into where the rock was compact," he wrote in his autobiography.

In the rock, Cope found, for the first time in America, a specimen of a horned dinosaur. He named it *Monoclonius*. Sternberg himself found two new species, including *Monoclonius sphenocerus*, another horned dinosaur which Cope estimated to have been 25 feet long and 6 or 7 feet high at the hips.

Sternberg spent many years collecting for Cope. The two remained friends until the professor's death in 1897. That friendship, however, did not prevent Sternberg from collecting for Marsh after Cope's financial ruin. He also collected for and sold his fossils to numerous museums around the world.

In the summer of 1901, Sternberg journeyed to the red beds of Texas, near Willow Springs, for the Royal Museum of Munich. By that time, fossil collecting had become a family affair. Sternberg's sons, George, Levi, and Charlie, all worked with their father in the field. The weather was so hot, "I saw cattle die of thirst and starvation," Sternberg wrote. "How can I describe the hot winds, carrying on their wings clouds of dust, which were so common that year and the next?"

Again, success mitigated the hardship. Sternberg found a relatively complete skeleton of *Labidosaurus,* an ancient and primitive reptile. The specimen became a prize of the Munich collection.

Over the next few years, the Sternbergs continued to work in Texas and the Badlands. In 1908, they went to southern Wyoming to look for *Triceratops*. On the same day they found a *Triceratops* skull, they found other dinosaur bones weathering out of a sandstone escarpment. After spending several days excavating the sandstone, they took their chisels to a block containing the breastbone.

Sternberg, who had gone to town to ship the *Triceratops* skull to the British Museum, recounts what he saw upon his return:

> Shall I ever experience such joy as when I stood in the quarry
> for the first time, and beheld lying in state the most complete
> skeleton of an extinct animal I have ever seen, after 40 years of

experience as a collector! The crowning specimen of my life's work!

A great duck-billed dinosaur . . . lay on its back with front limbs stretched out as if imploring aid, while the hind limbs in a convulsive effort were drawn up and folded against the walls of the abdomen. The head lay under the right shoulder.

Sternberg speculated that the dinosaur may have fallen on its back into a morass and died of a broken neck or drowned. "If this was so, the antiseptic character of the peat-bog had preserved the flesh, until, through decay, the contents of the viscera had been replaced with sand. It lay there with expanded ribs as in life, wrapped in the impressions of the skin whose beautiful patterns of octagonal plates marked the fine sandstone above the bones."

Why was the Sternbergs' duck-bill, which to this day is known as "the mummy dinosaur," so well preserved? After dying in the water, "the gases forming in the body floated the carcass," Sternberg surmised. "When the gases escaped, the skin collapsed and occupied their space."

Sternberg sold this find of his life to the American Museum of Natural History, and whenever he was in New York, he made a special point of seeing it. He continued collecting into his eighties and died at age 93 in 1943. His sons also continued to collect and were responsible for numerous important dinosaur finds, primarily in Canada.

Looking back on his years as a fossil hunter, Sternberg wrote:

What is it that urges a man to risk his life in these precipitous fossil beds? I can only answer for myself, but with me there were two motives, the desire to add to human knowledge, which has been the greatest motive of my life, and the hunting instinct, which is deeply planted in my heart. Not the desire to destroy life, but to see it. The man whose love for wild animals is most deeply developed is not he who ruthlessly takes their lives but . . . he who studies them with loving sympathy and pictures them in their various haunts. It is thus that I love creatures of other ages and that I want to become acquainted with them in their natural environments. They are never dead to me; my imagination breathes life into the "valley of dry bones," and not only do the living forms of the animals stand before me, but the countries which they inhabited rise for me through the mist of the ages.

Charles H. Sternberg said he was only answering for himself, but his lyrical reflection might just as well have been written by Peter Larson. In fact, Sternberg was one of Larson's heroes. The two shared more than a desire to work independently and think imaginatively. Sternberg also had experienced the loss of a great find.

While collecting in Alberta for the British Museum in June 1916, Sternberg and his sons had found three duck-billed dinosaurs. Two of the specimens were almost complete. After four months of excavation, Sternberg loaded the two extraordinary fossils onto the England-bound SS *Mount Temple*. The ship never made it across the Atlantic. Torpedoed by the Germans, it sank to the ocean floor, a casualty of World War I. Sternberg's heart sank with the vessel: "The British Museum could have mounted these two lords of the ancient bayou in that great storehouse of treasures, more rare than gold or silver, to be the heritage of the ages to come."

There was no indication that Sue would soon be mounted for the coming ages. Although Judge Battey quickly denied Duffy's motion for a rehearing in his court, the Eighth Circuit was obliged to hear the institute's appeal of the decision denying it ownership. This appellate process was certain to take months, if not years. The parties needed time to write their briefs and their replies to each others briefs. Oral arguments before the court would eventually follow. When the Eighth Circuit finally did render an opinion, the losing party might appeal to the U.S. Supreme Court. Conceivably one of the courts might order Judge Battey to hold a full-fledged trial to determine ownership.

Following Battey's February 3 decision, Schieffer had asked the Clinton administration to determine who owned Sue. South Dakota Governor George Mickelson, fearful that Sue might end up outside his state, had also asked Bruce Babbitt, the new secretary of the U.S. Department of the Interior, to review the matter. The administration might rule before the Eighth Circuit issued an opinion. But until the judicial branch was finally done with the case—whenever that might be—enforcement of any executive branch decision would, most likely, wait.

Schieffer would not see the case to its conclusion. In March, Senator Daschle announced that he had nominated Karen Schreier, an attorney from Sioux Falls, to replace the acting U.S. attorney. Senate confirmation was required, but Daschle suggested that Schreier might assume the office in an "interim" capacity, subject to confirmation at a later date.

Reached by the *Journal* after her nomination, Schreier appeared to be one of the few people in South Dakota who refused to answer: Who owns Sue? "I would have to look at the case, talk with the attorneys that have been handling the matter before I could make any decision," she told the newspaper.

Duffy's response to news of the appointment was injudicious. "It's hard to say at this point how things will play out, but, God only knows, nothing could be worse for my clients than Kevin Schieffer," said the institute attorney. "If Tom [Daschle] had announced that he was appointing Idi Amin, I would be dancing a jig. Unlike Schieffer, she [Schreier] is actually a lawyer and has some legal experience."

Of course, Schieffer did have a law degree and, for better or worse, had bested Duffy and his law firm at every turn. More important, until he left the post, Schieffer was still in charge of the investigation that could result in the indictment and imprisonment of Duffy's clients. Comparing the man who still had the most influence with the grand jury weighing the Larsons' fate to Idi Amin seems not only impolite but impolitic.

And "unprofessional," according to Schieffer. "If another attorney wants to make unprofessional comments, then that's their prerogative," he responded. "But I'm not going to . . . wallow in the mud on a professional level, and I'm not going to comment on the specifics of any cases."

When asked if criminal charges would be filed, Schieffer didn't wallow. Still, he couldn't resist returning Duffy's fire. "Theoretically, I'm not supposed to be in the business of confirming that there is a case," he said. "The only reason you guys [the press] know there is a case or all this concern for reputation . . . is because those folks [the institute] have been sending press releases to you right and left."

A few days after the Larsons' lawyer insulted Schieffer, one of their supporters announced plans to arrest the U.S. attorney . . . and Assistant U.S. Attorney David Zuercher . . . and FBI Special Agent

William Asbury . . . and Judge Battey. Travis Opdyke was fed up with the investigation. He felt the government had harassed his wife Marion Zenker, the institute's administrative assistant, when she appeared before the grand jury. Zenker had already incurred $8000 in legal fees.

Opdyke, a former deputy and jailer turned freelance writer, had tried to remove the prosecutors from their posts through conventional channels by forming the Government Accountability Group and petitioning the Justice Department and President Clinton. Having received no satisfaction from Washington, D.C., he was now ready to take matters into his own hands by arresting the four offenders. He emphasized that he would not be armed and would not use force in making the arrests. He urged Schieffer, Zuercher, Asbury, and Battey to turn themselves over to a magistrate "of their own accord."

Two days later Opdyke delivered a "notice of intent to apprehend" and a "bill of particulars" to Assistant U.S. Attorney Bob Mandel at the federal building in Rapid City. After Mandel convinced him that he would be "ill advised" to arrest a federal judge, Opdyke dropped Battey from his list and said he would work to impeach the judge. He gave the other three men 24 hours to turn themselves in before he arrested them. Contacted by the *Journal*, Larson said that he understood Opdyke's and Zenker's frustration but did not support Opdyke's latest plan. "These tactics do nothing to help our case," he said. Mandel told the *Journal* that the documents had no legal significance. His argument was moot. Opdyke never tried to make the arrests.

Had Opdyke known that, on March 18, Timothy S. Elliott, the acting solicitor of the U.S. Department of the Interior, had issued an opinion concerning the ownership of Sue, he might have tried to arrest him, too. The opinion, contained in a letter from Elliott to Governor Mickelson was not made public until April 7.

Who owns Sue? Maurice Williams, said the Interior Department.

Elliott's logic was simple. Williams had put his land in trust with the United States and couldn't sell or convey an interest in it without the Interior Department's consent. Judge Battey had ruled that the institute never received title to Sue because the Interior Department never approved the alleged transaction with Williams. Therefore, the land/fossil continued to be held in trust for Williams by the United States.

Elliott went on to discuss the implications of this judgment: "If the Indian landowner should decide to transfer ownership of the fossil, the proposed transfer would require approval by the Interior Department in its capacity as trustee. The sole purpose of the department's reviews, however, will be to ensure that the transaction is in the interest of the Indian landowner [Williams]."

Elliott did not mention that the Antiquities Act or any other federal statute might limit Williams's ability to transfer ownership. The department appeared to be concerned only that a transfer was in the rancher's interest—not the interest of the government or, for that matter the Cheyenne River Sioux. Presumably, Williams could sell Sue to whomever he pleased.

The press immediately sought reaction to Elliott's letter from the ubiquitous quartet of talking heads.

Schieffer chose not to respond. Two days later he announced that he was resigning immediately to enter private practice in Sioux Falls. On leaving, he said that he did not envy the "political and other pressure" his successor would face in dealing with the criminal investigation of the institute.

Years later, Schieffer was willing to share his thoughts. Awarding custody to Williams was, he said, "the least desirable" option. "I had gone around with the Department of the Interior on how it should be handled to prevent that kind of conclusion." He added that, at least in the court of public opinion, "The tribe was screwed. They had a claim. This was reservation land within their jurisdiction. Their rights from a civil action standpoint were not [understood]. The tribal chief tried to do the right thing. This was not about money but about sovereignty."

Despite his disappointment in the Interior Department's decision, Schieffer expressed a sort of grudging admiration for Williams's savvy. "The institute and Williams were both trying to do the same to the other. Who really got bamboozled?" Schieffer suggests it was the institute because, from the beginning, "Williams knew much more about Indian trust law than he let on." Had Williams suckered the Larsons from day one, taking their money when he knew all along that trust law prevented him from selling the bones? If so, he is not talking.

Williams was happy to respond to the Department of the Interior's decision. "It's real refreshing that we have laws," he said of the ruling that might eventually bring him financial reward.

Duffy continued his offensive offensive. "It's shocking to me that Kevin Schieffer has spent millions of dollars to allow Maurice Williams to stick a priceless paleontological object in a pole barn outside Eagle Butte, if he chooses to do so," he said.

Indian Country Today, "America's Indian Newspaper," quickly responded to these remarks. "Pat Duffy . . . is probably the poorest loser in legal history," the paper editorialized. "His latest rantings have even besmirched the sovereign Cheyenne River Sioux tribe. . . . Technically and historically, the *T. rex* bones belong to the [tribe]. If they work out an agreement with Mr. Williams it is of no concern—and certainly no business—of Mr. Duffy. To use such an archaic, paternalistic attitude toward the Lakota people of the Cheyenne River Reservation is appalling."

Duffy didn't back down. He responded to the editorial with a letter that the paper published. "My comment was aimed at a policy which has squandered millions and used the criminal law to seize a priceless paleontological relic for an individual who actually sold it (not, as you represent, for a tribe or an entire people)," he explained. "The personal comments about me, however, stretch facts . . . to find racism in my remarks, which only tangentially concerned a rancher who happens to be of Indian descent. My experience has been that those who stretch to reach overtly racist conclusions are often reacting from a racist perspective themselves."

Indian Country Today had suggested that the Cheyenne River Sioux owned Sue. The tribe responded to the Interior Department's letter by suing the institute, the Larsons, Bob Farrar, and Maurice Williams in tribal court. The suit sought possession of Sue and monetary damages for other fossils illegally removed from the reservation and no longer in the possession of the defendants.

On what grounds was the tribe claiming ownership? The suit alleged that Williams violated a tribal ordinance by failing to obtain a $100 tribal business license before allowing the fossil hunters onto his property or notifying the tribe of the hunters' intentions.

On what grounds was giving Sue to the tribe the proper remedy to what appeared to be a minor violation? A second tribal ordinance, enacted more than two years after Sue's discovery, sanctioned forfeiture to the tribe of "all property of any description . . . all moneys, negotiable instruments, securities or other things of value" used in violation of any tribal law. A *T. rex* was an "other thing of value."

For once Duffy and Williams agreed on something. Each said that the suit was ridiculous. Duffy accused the tribe of declaring war on one of its own members by challenging the ownership of property held in trust by the federal government. He also told the *Timber Lake Topic,* "If this case would stand up, it would set economic development back 100 years." Linking the license to forfeiture "will scare the hell out of anybody trying to do business in Indian country," he explained.

Williams said that if he needed a license to sell a fossil, then people picking berries, digging turnips, or selling eggs needed licenses, too. "They're [the tribe] running plain wild. This is going to progress into a hell of a lawsuit," he said. He added that he would rely on the federal government to protect his rights. Such protection appeared forthcoming. Bob Walker, a spokesman for the Department of the Interior, told the *Journal* that the department still believed Sue belonged to Williams. "I would think that in any subsequent legal action we would seek to intervene to uphold that principle."

Duffy and his law partner Mark Marshall questioned that principle in their appeal to the Eighth Circuit—the court that had the power to reverse Judge Battey's decision awarding ownership of Sue to the United States as Williams's trustee. In their appellate brief, filed just a few days after the tribal court action, Duffy and Marshall argued that Battey had erred in concluding that the government, as Williams' trustee, owned Sue because she was land or an interest in land. A partially embedded fossil is personal property, not real property, the institute's lawyers again maintained.

NATURAL HISTORY FESTIVAL
SCHEDULE OF EVENTS
SUNDAY, MAY 16

Bike-A-Thon	11:00	Football Field
Rock Hound Roundup	11:30	Black Hills Institute
Girl Scout Cookie Sale	All day	Around town
Open House	11:30	B. H. Institute
Music	11:30	Alpine Inn
Book Signing & Sale	12:00	Oriana's

Presentation	12:00	B. H. Institute
Art Print Unveiling	1:00	School Gym
Auction	1:00	School Gym
Bus Rides to Site	3:00	School Gym
Site Dedication	3:45	Museum Site
Potluck Dinner	5:00	Senior Center

The residents of Hill City had hoped that the discovery of Sue would reinvigorate their depressed economy. But instead of attracting crowds of tourists, the *T. rex* had attracted crowds of FBI agents and National Guardsmen. As the first anniversary of the seizure approached, only the lawyers had reaped economic benefit. However, for one day, anyway, the institute, the Larsons, and the entire town would forget the disappointments of the last 12 months, including the layoff of 44 mill workers and loggers by Continental Lumber Company, the town's largest employer.

Instead of gathering in district court or tribal court or before the Eighth Circuit or in the grand jury room, everyone would meet on the football field or in the school gym or the senior center or at the site of the new Black Hills Museum of Natural History.

Despite the disappointments of the last year, the town had decided that a celebration was in order. Mayor Drue Vitter had declared May 16 Hill City's First Annual Natural History Day Festival. The institute wholeheartedly endorsed the idea. "This could have been a dark day for us," said Larson, "but instead, on the anniversary of Sue's seizure, we want to affirm that we are definitely alive and kicking."

The seizure had produced one unexpected dividend for Larson. In December, Kristin Donnan, a writer who worked on the television show *Unsolved Mysteries,* had come to Hill City from Los Angeles looking for a story. Instead, Donnan, an attractive woman in her thirties with long brown hair, found love—during her first interview with Larson. A whirl-wind courtship had ensued and the two had married four months later.

There was other good news besides the marriage. Over the last 12 months, the Larson brothers had moved closer to realizing their lifelong dream of building a natural history museum. Their not-for-profit corporation had contracted to purchase a 10-acre site for the museum for

$50,000. Event organizers hoped that the day's events would raise the $7000 needed to make the first year's contract payment on the property. They exceeded that figure by almost $3000.

Almost 500 people turned out for the festival. Hill City youngsters began the day with a bike-a-thon that raised more than $150. That event's top fund raiser, Arlo Holsworth, received a cast of a *T. rex* tooth as his prize. An all-day bake sale raised additional money, as did the sale of a limited edition of 300 color prints of artist Mitch McClain's rendition of the museum. Casey Derflinger bought print number 1 for $250 and promptly displayed it in his office at First Western Bank. As expected, an auction of fossils and jewelry raised the most money, almost $9000. Among the 119 items sold were a cast of an adolescent mastodon skull with tusks, a cast of the world's largest shark's tooth, and a cast of one of Stan's teeth (Stan the *T. rex*, not Stan Sacrison).

Other activities contributed to creating a pleasant time-out from the Contentious Period. The popular Rushmore German Band from Rapid City provided music on the front porch of the Alpine Inn. According to the *Hill City Prevailer,* "The [potluck] supper crowd filled the Senior Citizens Center three times over as the line kept going and going and going. . . . No one went away hungry."

The highlight of the festival was the museum site dedication ceremony on a hill north of town. The Boy Scouts raised the flag. The Fife and Drum Corps gave their first show of the season. And Larson broke ground by digging a shovel full of dirt. "Pete and Neal gave it a quick lookover for any potential fossils it might contain!" reported the *Prevailer.*

The Larsons also spoke at the dedication. They reiterated the underlying philosophy of their efforts: that rather than being storage space for old fossils, a museum should be for the living—a place to uplift and educate. Through all the turmoil of the previous year, the Larsons had never stopped their education outreach programs. Most popular was their Dinosaurs on Wheels program, which featured a juvenile duckbilled dinosaur that had been specially mounted to facilitate its portability. The Larsons took Dinosaurs on Wheels to numerous schools around the state, including schools on Indian reservations.

When, in mid-May, Larson brought the exhibit to the grade school he had attended as a child in Mission, South Dakota, the local *Todd County Tribune* covered the two-day event:

Complete with a dinosaur skeleton, Larson's presentation fascinated the youngsters, who asked such questions as, "How many eggs do they lay?" and "Why are their arms so short in front?" In his talk, Larson explained that the creatures were really birds, told about methods of collecting and cleaning the fossils, and about what scientists do to learn about dinosaurs. "I love for them to ask questions," he said. "The only stupid question is the one you don't ask." He added that because so little is known about dinosaurs and the fossils are so plentiful, anyone can discover and contribute to paleontology.

The time-out ended about three weeks later. "Sue II: Hill City Ire and Federal Agents Return," the *Journal* reported on June 8. At 7:30 the previous morning, some 30 FBI agents, again led by Charles Draper and William Asbury, had descended on the institute with a search warrant. The agents had spent seven hours collecting fossils and documents, loading them into more than 50 boxes, and carrying everything out to two rental moving trucks. Ted McBride, interim U.S. attorney following Schieffer's resignation, explained that the raid was part of an "ongoing investigation."

Again, a portion of the institute was cordoned off with yellow evidence tape. And again Hill City residents rushed down Main Street to protest. *Journal* reporter Hugh O' Gara noted that by 10:00 AM, the nearby Heart of the Hills Convenience Store was out of cardboard delivery boxes. "Residents had commandeered the boxes to fashion impromptu protest signs."

"IS THIS AN ANNUAL EVENT?" read one sign. "YOUR TAX DOLLARS AT WORK," read another.

O'Gara reported the following conversation between two of the protesters:

"How do you spell 'seizure'?"

"You would think you would know by now."

Duffy was out of town for the raid. On his return he claimed that the raid was illegal. He called the search warrant "Jurassic Farce," a play on the title of the just-released blockbuster movie about dinosaurs, *Jurassic Park*. Continuing the motion picture theme, he called the agents the "Keystone Cops of paleontology."

Larson, who had been in Japan on business when the FBI came, could only scratch his head. "I can't understand why the Department of Justice feels it necessary to spend all this money simply to destroy our business," he said.

Mayor Vitter was equally confused. Were the Larsons and Farrar, as most people in Hill City thought, scientist-entrepreneurs who were committed to uplifting and educating while making a living? Or were they felons? At times it seemed that the U.S. attorney was treating them as if they were mafiosi running some sinister criminal enterprise with the institute as a front. Larson seemed like a nice guy with his visits to schools, but, then again, Don Corleone had been a loving grandfather as well as a godfather.

"I feel either we have the three worst criminals in South Dakota or the federal government is wasting its time," the mayor said following the raid. "It would be best for the town of Hill City to either charge these people and get it done or leave the case alone."

8

YOU CAN INDICT A
HAM SANDWICH

"Four million eight hundred thousand."

Unlike Peter Larson, Dr. Dale Russell was not amazed by the numbers. He and his colleagues from the North Carolina State Museum of Natural History had come to New York prepared to spend considerably more to win Sue.

No one had ever paid anywhere near this much for a fossil. Why wasn't Russell surprised? The specimen was superb, he would later explain. But there was more than that: "All the attention. All the lore. All the stories. Even though they aren't happy, they are stories. [They made Sue] like poor King Tut in his tomb."

While the government continued to search for fossils at the institute, Peter Larson continued to search for them in the field and through, for lack of a better term, his field representatives. The affidavits supporting the government's search warrants alleged that the Larsons were operating a multistate criminal enterprise. The institute's activities did extend beyond the borders of South Dakota. Institute crews conducted digs in nearby states as well as South America. Specimens were sold to institutions around the country and around the world. And Larson periodically traveled across state lines to buy fossils from amateur and professional collectors. Was there any criminal activity involved in Larson's "collection" business?

Larson didn't seem to be hiding anything. In the midst of the government's investigation, he invited journalist Greg Breining to accompany him on one of his multistate business trips. Breining later chronicled the trip in an article for *Wyoming Wildlife* magazine. On a wintry day, reporter and subject climbed into Larson's rusting Chevy Suburban and headed west. They sped across the open highway into Wyoming, where Larson pointed out the Morrison Formation, in which Othniel C. Marsh had found brontosaurs, and the Lance Creek Formation, where one of Larson's heroes, John Bell Hatcher, enjoyed remarkable success in the late nineteenth century.

Hatcher grew up in Iowa. To earn money for college, he worked in a coal mine. Fossils were plentiful there, and he soon became fascinated with things paleontological. After graduating from Yale in 1884, he asked Professor Marsh for a job. Marsh sent him to Kansas to work with Charles H. Sternberg. Soon, however, Hatcher became an independent collector for Marsh.

Hatcher is perhaps best known for his discoveries of the skulls of the horned dinosaurs, such as *Triceratops*. He found the first such skull on record while digging in Montana in 1888. In Wyoming the next year, Hatcher ran across a rancher who told him of a strange, huge, horned skull he had found. The rancher had tried to lasso the fossil and drag it back to his home, but the horn had broken off. Hatcher sent the horn to Marsh in New Haven and promptly received instructions to find the rest of the skull. He did, and the fossil became the type specimen of the dinosaur Marsh called *Triceratops horridus* ("terrifying three-horned face"). From 1889 to 1892, Hatcher mined the Wyoming formations for more dinosaurs. He found about 50 horned skulls in all, most of them complete.

Hatcher could find the small as well as the large. Near Lance Creek he found over 800 teeth of small mammals of the Cretaceous period. On one day alone, he found 87 teeth as he sifted through the dirt.

The sheer number of specimens wowed scientists. Some years later Hatcher revealed the secret of his success: He looked for the small fossils near anthills. Red ants apparently cleared the bones out of the way before building their homes. They then stacked them (as well as gravel) on top of the hills, perhaps to protect against the rain or wind.

Like many of Marsh's collectors, Hatcher eventually grew unhappy with their business arrangement. He left Wyoming for New Jersey in 1893 to become curator of vertebrate paleontology at Princeton. Later, after a stint at the Carnegie Museum in Pittsburgh, he went looking for more prehistoric mammals in South America.

Wilford calls the expedition "the stuff of which legends are made—and Hatcher is legendary in vertebrate paleontology." The Hatcher party once spent five months in the hinterlands without seeing anyone else. When money ran out, Hatcher would venture into the closest town and find or start a game of poker. Invariably, he won enough money to keep the expedition afloat. The locals didn't always appreciate his success at the table. After one particularly fruitful evening, Hatcher had to draw his gun to make a safe exit with his winnings.

Hatcher eventually returned to the United States. In 1904 he went west to finish a book on the horned dinosaurs. There he contracted typhoid fever from which he never recovered. He died that year at the age of 43.

Larson did not expect to find any horned dinosaurs on his trip to Wyoming. Instead, bats and birds were on his wish list. After several hours on the road, he stopped in Rock Springs. There, he hoped to buy a rare bat from a man and his sons who collected in the sediments of the Green River. When Breining looked at the specimen, he observed, "The bat is simply the hint of a shape in a slab of limestone, a little lump barely larger than a walnut. Few people would think anything of it at all." But Larson had.

When first shown the fossil a year earlier, Larson had borrowed it and arranged for X rays. The wings were clearly visible in the negatives. Unfortunately, on this day, the owner wanted more for the bat than Larson was willing to pay. Larson and Breining left empty-handed.

In Thayne, Wyoming, Larson called on a husband and wife team that also collected in the Green River sediments. The couple owned a rock shop and sold fossils to museums around the world, including the Smithsonian. They showed Larson another slab of limestone, this one containing a fossil of a bird. Again, the fossil was barely visible. When Larson held it up to the light, however, Breining observed that "shadows accentuate the wispy edges of feathers." Larson told the reporter that without skilled collectors like this couple, scientists would never get the opportunity to see numerous quality fossils like the bird. Unable to strike a deal for the bird, Larson drove on.

In Kemmerer, Wyoming, Larson did acquire a specimen—at least temporarily. A collector there had found a large slab with the remains of a creature the size of a small cat. Larson asked him where he had found it. On private land with the permission of the owner, the finder told him. Assured that it had been collected legally, Larson agreed to take the piece back to Hill City, where he would remove the skeleton from the rock and try to identify it. At that point, he and the finder would try to agree on a purchase price.

Before leaving Kemmerer, Larson also paid a visit to Ulrich's Fish Fossil Gallery. Breining reported that tension existed between the Ulrichs and other collectors in the area because the Ulrichs opposed liberalization of regulations governing commercial collecting on private lands. "I don't agree that any of the rare species should be sold," Wallace Ulrich told Breining. "They are the property of the schoolchildren." Breining wrote that Larson "snorted in disgust" after leaving the gallery.

The collector who gave Larson the slab to take home was subsequently arrested for and pleaded guilty to misdemeanor theft of government property for collecting "minifish" on BLM property. He was fined $1025. Larson had nothing to do with this incident, but it demonstrated the danger inherent in his line of work and also in his way of conducting business. As a middleman, he ran the risk of acquiring illegally collected fossils. This risk was compounded by his willingness to accept the word of those selling him specimens. He trusted them to be honest when they told him where they found their fossils. It remained to be seen whether the government would hold him responsible for unknowingly purchasing illegally collected fossils and then reselling them or would allege that he had done so knowingly.

Bakker describes Larson as a "contemporary Billy Budd, totally without guile." The Larson that Bakker, Hendrickson, and many others knew refused to see the potential for malfeasance or even misfeasance in others. He knew his word was good, so he assumed the word of others was good as well. As a result, he almost always sealed deals with a handshake rather than a contract. This practice did not always serve him well, as the failed deal for the duck-bill with the Viennese museum and the debacle with Maurice Williams so clearly demonstrated.

Larson's apparent faith in the goodness of his fellow humans was usually well placed. The painful experience with Sue was the exception, while the joyous experience with Stan was the rule. Thus, when Stan Sacrison called a few weeks after the Sue II raid to say he'd discovered some bones, Larson hopped in his truck and headed to Buffalo. Soon he had an announcement.

"We welcome back Duffy to the land of the living," Larson told the *Journal*. No, Patrick Duffy had not died and come back to life. This Duffy was Sacrison's latest find, a *Tyrannosaurus rex* whom Larson named in honor of the institute's "tenacious" lawyer.

Over the Fourth of July weekend, Sacrison had been hunting dinosaurs on the private ranch where he had found Stan. There, on top of a steep butte, he had spied some ribs and vertebrae and a pelvic bone within shouting distance of the cliff where he had discovered the dinosaur who bore his name. Certain this time that the bones were not *Triceratops*, he had quickly called Larson. Within a week, a crew of 12 was excavating the institute's third *T. rex* in just over two years.

Larson noted that the institute's good fortune in finding Sue, Stan, and now Duffy suggested that *T. rex* may not have been so rare as some scientists speculated. He added that if the institute could afford to send a crew into the field for an entire summer, it could find a new *T. rex* every year.

Heavy thunderstorms hampered the crew, as did the mud the storms beget. Still, after about a week, Neal Larson announced that they had already found about 25 percent of the skeleton, including some skull bones, half of its backbone, and both of its shoulder blades. Duffy, therefore, qualified as a major find—the fifteenth *T. rex* ever unearthed. Peter Larson was quick to point out that only one of these significant discoveries had been made by a degreed paleontologist, a not-so-subtle dig at the SVP's efforts to restrict amateurs and others associated with commercial collectors.

Duffy the dinosaur, like Duffy the lawyer, attracted a good deal of media attention. Writers from Holland and Scandinavia came to Buffalo. Japan's *Gakken Dinosaur Magazine* also sent a team led by Yoshio Ito of the National Science Museum to cover the excavation. This magazine had already published a story by Ito about Sue, as had magazines throughout Europe. "Sue ou l'Affaire du Tyrannosaure," read one French headline.

"Dinosaurier in 'Lebensgefahr': Rechsstreit um Besitzanspruche an 65 Millionen Jahre altem Fossil," trumpeted a German paper. "Steen des Aanstots uit de Prehistorie," reported a Dutch newspaper; a photograph of Wentz standing behind Sue's skull accompanied the article.

All manner of the American press was equally engaged. *Nova*, the award-winning Public Broadcasting System science show, sent a film crew to the dig, as did the less highbrow *Hard Copy*. Publications ranging from *National Geographic* to the *National Enquirer* also ran stories on dinosaurs.

"Grave Robbers Are Stealing America's Dinosaur Treasures," warned the *Enquirer*. The article that followed began: "The hit movie *Jurassic Park* has boosted public interest in dinosaurs to an all-time high—and has also spawned a multimillion dollar crime wave! Gun-toting dinosaur rustlers are illegally digging up and selling fossils to private collectors for up to $5 million apiece." Neither Sue nor the Larsons were mentioned, but the *Enquirer* quoted a paleontologist from the Utah State Bureau of Land Management who said, "It was rumored recently that a large carnivorous dinosaur was offered to a Japanese bank for $5 million." The *Enquirer* also reported the sale of a stegosaurus to a "Japanese interest" for more than $1 million. This was more than rumor. In July 1993 the Associated Press reported that a Utah company, Western Paleo Lab, acknowledged that it had shipped a complete stegosaurus skeleton to Asia. According to the company's Jeff Parker, "a group of individuals" had hired Western Paleo Lab to prepare the specimen, which had been found on private land in Wyoming. Parker said that a Japanese company had purchased it for a museum that it sponsored—the Hayashibara Museum in Okayama. Under the terms of the sale, the specimen was to be made available for scientific description.

Many American scientists were outraged by the sale. "It would be totally unthinkable for us to sell bones to Japan," Don Burge of the College of Eastern Utah's Prehistoric Museum told the AP.

The scientific community's criticism of the sale of one-of-a-kind fossils to private collectors at home or abroad was understandable; in such hands, these specimens might never be studied. But why the outcry if a fossil went from America to a foreign museum? For decades American museums had been digging and collecting abroad, often in third-world countries. In recent years, these museums had entered into contracts that assured those countries retention of ownership of the specimens. Still, the bones often ended up in the United States for years so they could be studied and, often, displayed. Should U.S. museums be criticized for robbing these nations of their national treasures?

"There's definitely a double standard here," says Robert Bakker. Why? "Can you spell 'jingoism'?" he asks. "The high priests of American paleontology see a 'yellow peril.'" Bakker notes that Jim Madsen, a Utah-based paleontologist, made sure his allosaur finds went to museums in several countries. "They're being studied in 20 languages," Bakker says. "And that's good."

In the meantime, tons and tons of bones discovered in America sit in museum or university storage facilities here, says Bakker. "The institutions simply don't have the funds to prepare them or the space to display them," Bakker explains. He notes that it takes about 20 days to prepare a specimen in the lab for every day spent in the field finding and excavating it.

The cost versus benefit of keeping a crew in Buffalo eventually forced Larson to make a tough decision. By the end of the second week at the site, the crew had not found Duffy's legs, pelvis, and large tail-bones. Most of the skull remained missing as well. The inability to find these bones was particularly frustrating because all signs pointed to their presence at the site. For example, the discovery of large teeth, complete with roots, indicated that the skull was somewhere nearby.

Larson reluctantly halted the dig. "The rest of the dinosaur could be 1 inch from where we stopped digging or 50 feet," he told the Associated Press. He stressed that he wasn't giving up. He hoped to enlist an oil company to provide special equipment that used electronic imagery to measure differences in the density of the soil. "Bone is generally less dense than rock," he said. "We just need to have something that gives us a direction to go in."

Larson had hoped that he could refine his theories about sexual dimorphism in *T. rex* by studying Duffy and Stan together. Duffy appeared to be a subadult and even smaller than the gracile Stan, whom Larson believed was a male. At this point, however, Duffy was too incomplete to shed much light on the subject.

Incomplete as he or she was, Duffy did shed light on Larson's theory about the social habits of *T. rex*. The fossil had been found within a quarter of a mile of Stan. The two dinosaurs' positions in the strata were similar, indicating that they could have lived at the same time. This appeared to support the theory that *T. rex* was not a solitary predator but rather lived in a family unit.

Duffy's discovery came at the same time that *Jurassic Park* was breaking box office records and triggering even greater interest in dinosaurs. In the movie, scientists use 65-million-year-old dinosaur DNA to bring back the long-extinct creatures. This raises the question: Could Larson have taken DNA samples from Stan and Duffy or from Sue and Dad and Junior and Baby to determine if they were actually family? "It is possible to get DNA from some of the bones," Larson says. "We had talked about doing that with Sue. There are some labs that would do it. But Sue was seized before we had a chance."

At the same time Larson told the Associated Press that he was temporarily abandoning the Duffy dig, he vented his frustration over the criminal investigation. "They're going to arrest us. They're going to bring charges in the next couple of months," Larson told the reporter whose name, perhaps fittingly, was Kafka (Joe, that is). "And then we'll have our day in court. Our fate will rest with 12 human beings at least."

Larson added that he had been optimistic that Karen Schreier's nomination as U.S. attorney would have put an end to the investigation. But the June raid had destroyed the hope that Kevin Schieffer's departure would stop the nonsense. "What's happened is they've spent so much money on this now that they can't back out," Larson continued. "They're out for blood. They've got to get us because their jobs are on the line."

Then for one of the few times in the 16 months since the seizure, Larson revealed his emotions in public. "I'm so tired of all this," he confessed. "It'll be a relief when they bring charges. This back room innuendo and all this other crap is really getting old."

The institute conducted about 20 percent of its business with Japanese museums, so perhaps it was inevitable that the innuendo would eventually extend to Asia. In late September 1993, Assistant U.S. Attorney Ted McBride, Laurie Bryant, a paleontologist for the Department of the Interior, and Douglas Rand, of the U.S. Customs Service flew to Japan. There they spent 11 days talking with institute clients and the Japanese authorities. The trip infuriated the Larsons and Duffy. Not only was it unnecessary, they argued, it was punitive—a frightening example of the government using its vast resources to destroy the little guy by intimidating those on whom he relies to conduct his business.

By chance Neal Larson had been in Japan setting up a dinosaur display at a museum when the government investigators arrived in Tokyo. He saw faxes regarding the investigation and contacted his brother back home. About the same time, a Japanese client sent Peter Larson a copy of a 17-page U.S. Justice Department memorandum introducing the McBride team to the Japanese Ministry of Justice. "Fossils, which are protected under the law as cultural objects, were stolen from America and Peru," the memo asserted. The Justice Department noted three suspect specimens sold by the institute to Japanese customers—a *Triceratops* found on the Cheyenne River Sioux Reservation, a *Triceratops* found on the Standing Rock Reservation, and a baleen whale found in Peru.

Duffy called the memo "a lie, a wicked fabrication" and then let loose a litany of pop culture references. "Robert Ludlum couldn't keep up with this conspiracy," he told the *Journal*. "There isn't a criminal mind short of Lex Luthor who could keep up with this." He added: "It's a perfect meld of *Jurassic Park* and *Rising Sun*. . . . What it boils down to is that Ted McBride thinks he's Wesley Snipes." In the newly released movie *Rising Sun*, actor Snipes played a Los Angeles cop who went to Japan to investigate a murder.

Duffy was incensed that someone had leaked the memo to one of the institute's customers. "If that isn't defamation, I don't know what is," he said. "My clients have never had the chance to defend themselves against this international smear campaign. It's paleontological McCarthyism, plain and simple. . . . This is the way people are put out of business by the Department of Justice." After being contacted by the McBride team, one Japanese client had already told Larson that it was wary of continuing to buy fossils from him with this cloud hanging overhead.

For the second time in a little more than a month, Larson publicly aired his anger. He told the Pierre *Capitol Journal* that there was no need for the McBride party's costly trip overseas. "We and the Japanese have already provided the U.S. government with information on all the specimens they are supposedly investigating in Japan," he said. "It's like they [the government] know we're bad. They know we're evil. They just want to destroy us any way they can. But they aren't charging us with anything because once they do, then we have rights—to see their allegations, to look at their alleged evidence. If we go down, we're going to go down fighting. But they're never going to destroy us."

With the Justice Department already under fire for its raid on the Branch Davidian compound in Waco, Texas, Duffy intoned: "I do not want Peter Larson to become the David Koresh of paleontology." He echoed his client's sentiments that it was time for the U.S. attorney to end the investigation and bring the inevitable indictments. "I just wish we could get them into court," Duffy said. "Let them bring their best lawyers from Washington—by the Greyhound busload if that's what they want. Right now we're stuck in a nether world."

So was Sue. Thus, on October 11, the three-judge panel from the Eighth Circuit heard oral arguments in the institute's appeal of Judge Battey's ownership ruling. Three years and two months had passed since Sue had been discovered. Eighteen months had passed since she had been seized and stored at the School of Mines. "Regardless of what happens, paleontologists on both sides agree that this is a tragedy," Duffy told the court.

Six weeks later the grand jury ended the institute's stay in the nether world. "Hill City Fossil Hunters Indicted," announced the *Journal* on the day before Thanksgiving, 1993.

The 33-page, 39-count indictment named the Black Hills Institute itself, Peter Larson, Neal Larson, Bob Farrar, and Terry Wentz as defendants. Eddie Cole, the Utah fossil collector, and his wife Ava were also charged for their alleged role in one transaction, as was a Californian with whom the institute had done business. One hundred fifty four separate offenses—148 felonies and 6 misdemeanors—were alleged. If convicted of all crimes, Peter Larson faced up to 353 years in prison and $13.35 million in fines.

What had Larson and his partners done? On its cover page, the indictment listed the following crimes in capital letters: "CONSPIRACY, ENTRY OF GOODS BY MEANS OF FALSE STATEMENT, THEFT OF GOVERNMENT PROPERTY, FALSE STATEMENTS, WIRE FRAUD, OBSTRUCTION OF JUSTICE, MONEY LAUNDERING, INTERSTATE TRANSPORTATION OF STOLEN GOODS, STRUCTURING and CMIR [Currency or Monetary Instruments Reporting] VIOLATION."

The fossil hunters may not have worn suits and ties to the office, but as Mandel would later say, "This was basically white-collar crime." The institute defendants were charged with 14 separate instances of illegally collecting fossils. But the majority of the alleged offenses were related to what the Larsons and their partners did after they collected the fossils. As portrayed by the government, the defendants engaged in a sophisticated conspiracy to commit fraudulent criminal activity that included everything from falsifying documents about where the fossils had been obtained to laundering money gained from the sale of fossils to violating customs regulations. The conspiracy extended from Hill City to Peru and Japan.

How many of these counts were related to Sue? None. Sue was not even mentioned in the indictment.

The irony of this was not lost on Larson. The government had justified seizing Sue because she constituted evidence necessary for its criminal investigation. In its first opinion the Eighth Circuit had clearly stated that the government didn't need the bones to carry out that investigation. Then the government had admitted that the Antiquities Act—the statute it had principally relied on in securing the initial warrant—did not apply to this case. Now the prosecutors themselves were acknowledging that there was no criminal wrongdoing with respect to the collection of the celebrated fossil. Larson could only wonder what would have happened if the government had never taken Sue. Would Williams or the Department of the Interior or the Cheyenne River Sioux tribe have filed a civil lawsuit to gain possession of her from the institute?

The fossils that were listed in the indictment included several *Triceratops*, duck-bills, mosasaurs, whales, turtles, ammonite, crinoids, and catfish. They had allegedly been collected illegally from public— federal, tribal, and state—and private lands in South Dakota, Wyoming, Montana, and Nebraska. Some of these fossils had been sold to the Smithsonian Institution and the Field Museum. Such illegal activity dated back to 1983, the indictment charged. A few examples:

- In 1984 and 1985, "principals, agents, or employees" of the institute allegedly went on U.S. Forest Service land—Gallatin National Forest—in Montana, and collected fossil remains of crinoids.

- In 1989, the Larson brothers allegedly went on Standing Rock Sioux Reservation land near Corson, South Dakota, and collected fossil remains of a *Triceratops*.

- In 1983, "employees or agents" of the institute allegedly went on private lands near Lantry, South Dakota, and obtained the fossil remains of various ammonites without the permission of the owner.

The Larsons had long maintained that they never knowingly engaged in the illegal collection, purchase, or sale of fossils; like Harvard's Dr. Jenkins, they didn't know if they were collecting on federal, state, or tribal land subject to restrictions. The government didn't believe this. The indictment charged that the defendants were adept at reading maps. As part of the conspiracy, however, the defendants and their coconspirators "would feign ignorance of property boundaries and property ownership in places they were illegally collecting fossils." In June 1989, for example, Peter Larson and Sue Hendrickson allegedly met with an Amoco employee near Wamsutter, Wyoming. This employee showed them maps that identified the location of BLM and private lands in the area. "Thereafter," the government alleged, "Peter Larson and Sue Hendrickson collected fossil remains, including a turtle from lands that were clearly marked as Bureau of Land Management lands on the map shown to them."

The conspiracy was alleged to have extended beyond fossils personally collected by institute personnel: "The defendants and coconspirators when purchasing fossils from other collectors would deliberately avoid gaining information regarding the location and details of illegal collection when they knew or suspected that the fossils were illegally collected."

The defendants' deviousness went beyond feigning ignorance. According to the indictment, "when dealing with others, [the defendants] would emphasize the educational and scientific benefits to be derived from their fossil-related activities and minimize or conceal the commercial benefit and personal economic gain to themselves." In 1988, for example, Larson allegedly collected the remains of a mastodon, camel, and three-toed horse from private land in Nebraska "by giving the landowners the false impression that the items were being collected for donation to universities and museums."

In addition to naive landowners, the defendants duped the government, universities, foreign corporations and museums, according to the government:

- Item: In 1987, the institute partners had imported a baleen whale fossil from Peru and "entered it into the commerce of the United States." They did so by means of a "false and fraudulent declaration [that] . . . the whale was of scientific value only and no commercial value," when they well knew that the whale was of "domestic commercial value" in excess of $10,000.

- Item: In 1992 (during the U.S. attorney's investigation), a company in Okayama, Japan, that was developing a natural history museum arranged to purchase a *Triceratops* skull from the institute for $125,000. One of the purchaser's requirements was that the institute issue a certificate that the fossil was excavated from private land and was, therefore, commercially tradable. The institute issued such a certificate despite the fact that the fossil was excavated from Sharkey Williams's land "held in trust by the United States for the Cheyenne River Sioux tribe." This was the specimen the institute was digging up on the property of Maurice Williams's brother when Sue was found.

The exhaustive investigation of the Larsons uncovered several alleged customs-related violations as well. Peter Larson was accused of failing to declare $31,700 in travelers checks carried from Japan to the United States. He was also charged with taking $15,000 in cash from Hill City to Peru without filling out the required United States Customs forms. The institute was also accused of knowingly undervaluing two fossil shipments to Japan.

Finally, the U.S. attorney did not forget Neal Larson's desperate, pre-raid attempt to change the dates on boxes of fossils. This, said Count XXXVII, was obstruction of justice. Another count accused both of the Larsons and Farrar with concealing or failing to produce records the government had subpoenaed.

The defendants were served the indictments on November 23, but they were not arrested. They were to be arraigned on December 15. Duffy, who represented the Larsons, Farrar, Wentz, and the institute, said his clients would plead not guilty. He demeaned the prosecution. "In this country you can indict a ham sandwich," he said.

As promised, at the appointed time the defendants pleaded not guilty to all charges. But December 15 was not without its surprise.

Purely by chance, on the same day the Larsons were arraigned, the Eighth Circuit ruled on the appeal of Judge Battey's decision that had denied the institute ownership (and possession) of Sue.

Peter Larson held his breath as he waited for Duffy to relay the court's opinion. Christmas was only ten days away. The U.S. attorney had already filled his stocking with fossil fuel (coal). The Eighth Circuit could go a long way to making this a happy holiday—in spite of the indictments—if it ruled that Sue should be home for Christmas.

9

NEGOTIATIONS ARE UNDER WAY

"Five million dollars."

Larson knew that the top paleontologists from several museums had visited Sotheby's before the auction to observe the condition of Sue's bones.

"Damn," he said. "They must think we did a terrific job with her." Maybe there was still a chance.

If a dinosaur falls on the land and nobody sees it, is it still a dinosaur?

Judge Magill said the Eighth Circuit had to make the following decision: "Whether the fossil was personal property or land before Black Hills excavated it." As federal statutes regulated Williams's property, the court looked to Congress for a definition of "land." Finding no applicable definition, Magill, like Battey, turned to South Dakota property law for guidance. He concluded:

We hold that the fossil was "land."... Sue Hendrickson found the fossil embedded in the land. Under South Dakota law, the fossil was an "ingredient" comprising part of the "solid material of the earth." It was a component part of Williams's land just like the soil, the rocks, and whatever other naturally occurring materials make up the earth of the ranch.... That the fossil was once a dinosaur which walked on the surface of the earth and that part of the fossil was protruding from the ground when

Hendrickson discovered it are irrelevant. The salient point is that the fossil had for millions of years been an "ingredient" of the earth that the United States holds in trust for Williams. The case might very well be different had someone found the fossil elsewhere and buried it in Williams's land or somehow inadvertently left it there. Here, however, a *Tyrannosaurus rex* died some 65 million years ago on what is now Indian trust land and its fossilized remains gradually became incorporated into that land. . . . We hold that the United States holds Sue in trust for Williams.

"What a vicious bit of irony that this decision comes on the day they [the Larsons] are arraigned on 39 counts," said Duffy. The institute would appeal the decision to the United States Supreme Court, he said.

Judge Battey set the criminal trial for February 22, 1994. It began on January 10, 1995. The 13 months following the arraignment were filled with public recriminations, private negotiations, and motion upon motion.

The institute fired the first shot. On January 6, 1994, three weeks after the arraignment, the Larson brothers each filed motions asking Judge Battey to recuse himself (step down) from the trial. They argued that his comments during the custody hearing months earlier suggested that he favored the prosecution and already believed the defendants guilty. In particular, the Larsons cited the judge's remarks that they had probably obstructed justice.

The government fired back the very next day. Assistant U.S. Attorney Zuercher filed a motion to disqualify Duffy from the trial. Zuercher asserted that Duffy had a conflict of interest because earlier in the case he represented other defendants and witnesses. Zuercher wanted Gary Colbath, who represented Sue Hendrickson, disqualified as well. Colbath was now representing the institute's Terry Wentz. The Larsons were furious. Peter Larson argued that the government was attempting to deny him his Sixth Amendment right to the attorney of his choice. Zuercher responded that the right to choose one's own attorney is not absolute.

Judge Battey refused to step down and refused to disqualify Duffy. Privately, some of Larson's supporters were disappointed that Duffy would remain. Hendrickson and others thought Duffy might not be the best lawyer for the task and that, at the very least, he needed some help from more seasoned practitioners. They reasoned that he had enjoyed little success to date and had alienated the government and the judge with several public pronouncements. Bill Mathers, a friend of Hendrickson, had found a prominent Washington, D.C., law firm that was apparently willing to undertake the defense pro bono publico ("for the good of the public"—at no charge to the client). "There were a lot of excellent attorneys who were outraged by the government's behavior," says Hendrickson. She believes that famous defense lawyer Gerry Spence might have been persuaded to help with the case if Larson had given permission to pursue him.

But Duffy didn't seem to be interested in outside assistance, and the intensely loyal Larson was not about to remove him. "Pat and I have been in this together since the beginning," he told the *Journal*'s Harlan. "I think he's the only one who can adequately defend me."

In mid-February, Duffy went back to court—this time in connection with the civil proceedings for possession of Sue. As the Supreme Court was under no obligation to hear an appeal and the odds of the court overturning the Eighth Circuit were slim anyway, Duffy decided that the time was right to get something in return for Sue. He therefore filed a mechanic's lien on the fossil. Such liens are generally filed by workmen who have not been paid for their labor on a piece of property. The institute's lien was for $209,000—about $35 per hour for 5630 hours spent finding, excavating, and preparing Sue. This case was eventually dismissed. So, too, was the Cheyenne River Sioux's suit for the *T. rex* in tribal court, bringing Williams one step closer to gaining Sue.

As the summer wore on, so, too, did pretrial motions in the criminal case. "Feds Want Fossil Trial in Aberdeen," announced the *Journal* on July 27. "Due to both the volume and nature of the publicity in this case, the United States argues that it would be difficult, if not impossible, to secure an unbiased panel of jurors in the Western Division," said Zuercher and Mandel. The Western Division included Rapid City, home of the *Journal*, which by the prosecutors' count had run at least 135 news stories about the case in the 26 months since the seizure. An exhibit accompanying the motion included editorials and letters to the editor that criticized the prosecutors' handling of the case.

Duffy saw the motion as another "misuse of prosecutorial power." The prosecution had estimated that the trial could take as long as three months. A move to Aberdeen, 300 miles to the northeast, would add $100,000 to the cost of the defense, said Larson's lawyer. Among other things, the move would necessitate getting motel rooms for half a dozen attorneys, an equal number of support staff, and the defendants. "It would stretch our supply lines to the breaking point," Duffy told the *Journal*.

"The prospect of living out of a motel in Aberdeen for ten weeks didn't exactly excite me," Assistant U.S. Attorney Mandel would later say. "It's not Honolulu." But, Mandel explained, the government truly feared that it would have a difficult time finding jurors who hadn't been influenced by the local press's coverage of the story. He noted that his office rarely spoke to *Journal* reporters and those reporters rarely called his office; as a result Duffy was often able to "spin" his side of the story without rebuttal.

In the days that followed the filing, the prosecutors may have wanted to add two more *Journal* pieces to their exhibit. "Open-minded people in the Rapid City area should be insulted," began an editorial in the paper titled "Trial Should Stay Here." The editorial noted that many of the 135 articles

> focused primarily on the government's confiscation of . . . Sue—even though none of the charges involve Sue. The implication is that the *Journal*'s coverage of the case has already convinced most people here that the accused are innocent. This is a dubious claim. . . . But even if it were true that potential jurors are already presuming the defendants' innocence, what's wrong with that? Isn't that what juries are supposed to do, to presume innocence and be convinced of guilt?

In a humorous column a few days later, Harlan observed, "Rapid City's soft on crime reputation is hurting our economy. For evidence look no further than that . . . prosecutors asked the judge to move the trial . . . to Aberdeen." He continued:

> Normally, this is the paragraph where I would explain what the . . . trial is about, but apparently I don't have to do that.

The feds say the fossil case has received so much publicity that everyone in Rapid City not only knows about it, they already have an opinion about it—namely that the defendants are not guilty. . . .

Which brings me back to money and why Rapid City needs to change its permissive image.

Justice Department officials won't say how much they have already spent in their three-year prosecution . . . but a total well into seven figures wouldn't surprise me. After all, investigators have been to Japan, South America, and Europe.

The trial itself . . . could last for months. It could involve witnesses from all over the country—including high-salaried government witnesses on lucrative per diems. The lunch trade alone could be worth thousands, and then there's dinner, motels, and legal pads.

Heck, this case is the economic development equivalent of an Airstream rally.

The economic development of Hill City remained on hold during this period. Some tourists visited the institute, but the lunch trade hardly was worth thousands. When Larson wasn't in the field, he was happy to show these visitors Stan's bones and Sue's skull. Not her real skull, but a bronze bust sculpted 1 : 8 scale by Joe Tippman, an artist from Hermosa.

In early August a second rendition of Sue's skull arrived in Hill City courtesy, strangely, of the very same arm of the government that had originally hauled off the real Sue—the South Dakota National Guard. The guard had been enlisted by Bancroft Elementary School in Sioux Falls. A year earlier second graders and 11 parents sympathetic to the institute had created a green, papier-mâché likeness of Sue's head. They wanted the Larsons to have something to remember the dinosaur if the courts ruled against them.

The likeness was a bit larger than Sue's real skull, and therein lay the problem. The school had been unable to find a means of transporting the copy to the institute until the guard volunteered. "It's almost like the National Guard did its penance," said the institute's Marion Zenker. "It's wonderful."

Even more wonderful was the discovery of another skull—Duffy's. In late August, Steve Sacrison had renewed the search for the bulk of the *T. rex*. About 40 feet from the area where the rest of the bones had been found, Sacrison discovered Duffy's lower jaw and part of the upper jaw. Larson announced that he expected they would find the rest of the skull. Then the commercial collector announced that Duffy was not for sale. The fossil, which Larson believed was a three-quarters grown subadult, would be displayed in the new museum with Stan—and Sue, if the Supreme Court ruled in the institute's favor.

Peter Larson was sitting in Patrick Duffy's office when the telephone call came. Hugh O'Gara, a staff writer for the ubiquitous *Journal*, was on to a big story. Was it true, O'Gara wanted to know, that a plea bargain in the criminal case was imminent? Larson remembers that Duffy confirmed the details of the negotiations.

On September 20, the *Journal* ran one of its biggest headlines since the discovery of Sue: "Fossil Case Won't Go to Trial." "Hit by several recent setbacks in the celebrated 'Sue' fossil prosecution, the federal government is virtually giving up the case," O'Gara wrote, citing "sources connected with the case." Larson has never been able to determine who those sources were.

At the time the trial was set to begin on November 1. But, wrote O'Gara: "Negotiations are under way that will result in the dismissal of all but one charge in the 28-month-old federal case. . . . Preliminary negotiations call for the institute—but none of the four principals . . . —to plead guilty to one felony charge that does not involve any of the fossils the government has claimed were illegally collected." The government would, however, keep all the disputed fossils it had seized from the institute.

This wasn't the first time the defendants and the government had discussed a plea bargain. But, according to Larson, up until this time prosecutors had always insisted that he, not his business, plead guilty to one felony and that he agree to 24 months in prison. "Those terms were unacceptable," according to the paleontologist, who maintained that he had done no wrong. As the trial approached, however, several developments led the U.S. attorney's office to soften its position and agree to let the institute itself take the fall. On September 6, Park Service Ranger

Stan Robins had died. The government considered him a key witness, as he had been among the first to investigate the institute.

Adverse rulings by Judge Battey also forced the prosecutors to reconsider the chances of success at trial and reopen negotiations with the defendants. On September 8, the judge had denied the government's motion to move the trial to Aberdeen. A Rapid City jury would, in all likelihood, be sympathetic to the defendants. O'Gara's exclusive article about a settlement ended rather curiously. "The plea agreement under negotiation calls for no publicity by either side in the settlement." Obviously, one of the sides had violated that term—for where else could O'Gara have received such detailed information?

Judge Battey apparently read about the plea bargain negotiations in the paper like everyone else. He didn't like it. At a pretrial hearing the day after O'Gara's story appeared, the judge spoke. The *Journal* secured a copy of the transcript of his remarks:

> From what I saw in the paper, if that's the plea agreement, it's not a plea agreement, it's a capitulation by the government. I am going to look very closely at whether or not in any plea agreement of your client, principal officers of a closed corporation can escape by putting fault over on the corporation. I made a decision on this.

Later the judge added:

> I don't want to enter into a plea agreement. . . . It looks to me from what I see in filings that there are relative degrees of fault, even among the conspiracy members. Some may be marginally involved; others may be highly involved. But I will wait. . . . The last thing lawyers should be worrying about now is a plea agreement.

Within a few days, Duffy was back in court—worrying about the judge. For the second time, the institute's lawyer filed a motion seeking that Battey recuse himself. Duffy cited the judge's recent comments as further evidence that the defendants were already presumed guilty by the man who would conduct their trial.

The government was back in court, too. The *Journal*'s article about the supposed plea bargain had once again compromised the government's ability to get an impartial jury in Rapid City, said prosecutors. They again moved that the trial be moved to Aberdeen. In documents filed to support the motion, Mandel asserted that the government had not offered a plea agreement. The *Journal* would later acknowledge this, stating that the article should not have given the impression that a tentative deal had been struck.

"There was no offer and no acceptance," Mandel explained in a 1999 interview. Negotiations were proceeding, the prosecutor said, but he refused to discuss the specifics. Larson says that reporter O'Gara "had it just about right." The deal, however, had not been sealed. When the story broke, Larson was trying to decide to which felony the institute would plead guilty. "I wasn't going to lie and say we'd done something we hadn't done," he explains.

Did the leak to the *Journal* prevent a deal? "If there had been an offer and acceptance, I don't think it would have mattered," Mandel speculated. Then he added: "Was it conducive to reaching an agreement? Well, there never was one after this."

Bob Chicoine, Hendrickson's lawyer, doesn't know who leaked the story, but he knows he would have handled O'Gara's call differently than Duffy did. "You say, 'No comment.' You don't discuss the status of plea negotiations with anyone—especially the press—until you have them buttoned up."

Neither the government nor Duffy prevailed on their motions. Judge Battey, furious with the *Journal* and whomever had leaked the story about the plea bargain, gave serious consideration to moving the trial to Aberdeen. In the end, however, he decided that both sides could get a fair trial in Rapid City. He also decided that the defendants could get a fair trial if he remained on the bench.

Running the risk of further alienating the judge, Duffy appealed this decision to the Eighth Circuit. A three-judge panel ruled unanimously that Battey need not step down; he had not shown the high degree of bias necessary for removal. "Judge Battey [performed] precisely the type of analysis that he is required to perform in analyzing a plea agreement, albeit prematurely," wrote the court.

One of the three judges did, however, take the time to write a separate opinion. Judge C. Arlen Beam worried that Battey's remarks might have "a chilling effect" on future plea bargain attempts. He added that he hoped a thaw might be forthcoming. "This is a matter in which the public interest may best be served by a plea bargain," he noted.

Many observers believed that the interest of the defendants would also best be served by a plea bargain and that one might have been achieved early on—long before the government and the defendants expended large sums of money. These same observers believe that the personalities and tactics of the principal attorneys made it difficult to effect such a bargain. "Would it have settled with different lawyers?" mused Mandel some years after the trial. "That might have been possible. Things were very polarized from the beginning and that doesn't help. In all my years as a lawyer, I have never seen a case played out in the press like this one."

Just as the Eighth Circuit ruled that Battey should stay where he was, the Supreme Court decided that Sue should remain where she was. The court denied the institute's petition to hear its appeal of the decision that, in effect, named Maurice Williams the fossil's rightful owner. Four years and two months after finding the *T. rex*, two years and six months after the seizure, Peter Larson had finally exhausted his legal remedies to win back the find of his lifetime.

Conventional love stories are played out in three acts. Act One: Boy meets girl and they fall in love. Act Two: Boy loses girl. Act Three: Boy gets girl back. Act Two in this most unconventional love story was now officially over.

Larson's ability to get Sue back now lay in the hands (or more properly the wallet) of Maurice Williams, not the judicial system. Williams still wasn't sure what he wanted to do with the tons of bones sitting at the School of Mines. He had received offers for Sue, but he wasn't certain that he wanted to sell her. "I had heard that the most money is taking it for tour," he told Harlan after the Supreme Court denied the institute's petition. One could picture beer companies vying for the rights to present the "Rolling Bones."

Sue was already a worldwide celebrity. The publicity attendant on the criminal trial of those who found her could only enhance her notoriety—and value. If, as Judge Beam said, the public interest would best be served by a plea bargain, Maurice Williams's interest was probably best served by a trial that would continue to keep Sue in the camera's glare.

As January 10, 1995, approached, it became clear that Williams would again be the winner. *United States v. Black Hills Institute, Peter Larson, et al.* was going to trial.

10

THEY'RE NOT CRIMES

"Five million dollars," Redden said a second time. The bidding had finally slowed down.

The three men in the private room above the floor had yet to be heard from. Dede Brooks rang them up. Were they planning on bidding? she asked.

The defendants moved down the street in a herd, followed by predators bearing minicams and note pads. They found refuge inside a three-story tall, white concrete structure—the federal building. But as they took their places in Judge Battey's large wood-paneled courtroom, a more dangerous threat presented itself—the two-headed prosecutorus lex, Bob Mandel and David Zuercher.

Mandel, an amiable man in his mid-forties with short, dark hair and a dark mustache, was born and raised in the big city—Chicago. He had emigrated to the Dakotas in the late 1970s after graduating from Antioch Law School in Washington, D.C. Initially, he worked for the Legal Services Corporation on an Indian reservation in North Dakota. In 1982 he joined the U.S. attorney's office in South Dakota. He had handled both civil and criminal cases and now supervised the Rapid City office. Although he understood why some people liked to collect fossils, he had no interest in doing so himself. "It's not my idea of entertainment," he explains.

The bearded Zuercher, in his mid-forties, was thinner and more temperamental than Mandel. A native of Nebraska, he had gone to college and law school in South Dakota. After working for the South Dakota attorney general, he had joined the U.S. attorney's office about 15 years earlier.

Judge Battey had been screening prospective jurors since the fall. Still the vast majority of the 34 men and women in the first panel to be questioned by the judge and the lawyers acknowledged that they had heard and read about the case. The prosecution did not aggressively attempt to dismiss those who were familiar with the events of the last two years and eight months. Mandel explains that it is very difficult to prove bias and that such an effort may antagonize those who are questioned or just observe the questioning—some of whom inevitably wind up on the jury. As a result, the government and defense quickly selected eleven women and three men to serve as jurors and alternates.

Mandel says that every time a case goes to trial, "you struggle with making it understandable for the jury." He admits that this was a particular challenge in this case as there were so many different instances of alleged illegal collecting and so many other alleged crimes involving financial transactions and customs violations. "You try to lay it all out for the jury in the indictment," the prosecutor says. The trial, therefore, began with Zuercher reading the full text of that 33-page document—a task that took more than an hour.

Mandel then began his opening statement. He promised the jurors that the evidence would show that the defendants engaged in a "far-reaching conspiracy for an extended period of time." He also promised: "This is not going to be a complicated case."

On its face, the case wasn't that complicated—at least with respect to fossil collecting. The jurors had merely to decide if the defendants had knowingly and willfully collected specimens from any or all of the federal, state, and tribal lands cited by the government. "In the abstract, you think, So what? Anyone can get screwed up [as to whether they are collecting on forbidden lands]," Mandel later explained. "My task was to try to focus on the fact that the defendants did have knowledge about what they were doing." That would require getting the jury to answer "yes" to two questions, Mandel says. "One: Did the defendants know it was illegal to collect on public lands? And two: Did they know they were on federal lands when they were collecting?"

Mandel told the jury that the evidence would show that the defendants had lobbied for laws permitting collecting on public lands. Therefore, they obviously knew that it was currently illegal to collect on those lands. The prosecutor also promised that the defendants' own field notes would prove that they knew they were on federal lands. They were very good map readers and map drawers, he said. In fact, government investigators following the defendants' notes and maps had found the sites and even the holes where fossils had been excavated.

Mandel showed the jurors a map of those sites. But the prosecutor couldn't present a map of the defendants' hearts and minds. And this was the map he needed. For it was clear from the outset that the fate of the defendants would rest on whether the jury viewed them as sophisticated criminals intent on ignoring the laws of the land or honest men who truly didn't know that they were doing anything wrong.

Honest men, proclaimed Duffy, who by this time represented only Peter Larson; the other defendants, including Neal Larson, had their own attorneys. To prepare for the trial, Duffy had gone to Montana, where, earlier, Gerry Spence had won a multimillion-dollar verdict in a civil lawsuit against the government following an infamous incident at Ruby Ridge, Idaho. Spence represented Randy Weaver, the white separatist whose wife and son had been killed in a shoot-out with the FBI. In Butte, Duffy went to the library and read all the newspaper clippings about the case. Then he talked to many residents to get a sense of how best to tap into or arouse the antigovernment sentiment of a jury.

Duffy's opening statement evoked memories of the "Man from Hope" video that in 1992 had introduced a different side of Bill Clinton to his jurors, the voters. The "Men from Hill City" presentation chronicled the Larson brothers' roots—their first museum, their dreams of collecting fossils—and their ultimate realization of their dreams, the American Dream. They now sold fossils to the greatest museums around the world, Duffy told the jury.

Having humanized the defendants, Duffy addressed the government's charges. So, too, did Neal Larson's lawyer, Bruce Ellison. The core of the Duffy/Ellison defense was simple: their clients never knowingly illegally collected on public lands. In some instances in which they may have collected on such land, they did so only after receiving permission from those whom they thought owned the land; in other words, they thought they were on private, not public, property. In other instances, they relied on the word of private collectors whom they trusted. And on still other occasions, they did not believe it was against the law to collect the particular type of specimens they were charged with stealing. In short: these honorable scientists either did no wrong or meant to do no wrong, said their lawyers.

For the next six weeks, Mandel and his team reconstructed their case as Peter Larson and his team would construct a dinosaur—bone by bone. Quite frankly, few visitors to the institute would find it exciting to watch Larson or Terry Wentz use their airbrades and epoxy for more than a few minutes. Similarly, few visitors to Judge Battey's courtroom were held spellbound by the government's use of 92 witnesses and hundreds of exhibits. "This was the kind of case where you bust your butt trying to prove the elements, subjecting the jury to tedious testimony," Mandel acknowledges.

Of course, in building a case or a dinosaur, it is the end product that is important. And, in fairness to the lawyers, there were more than a few telling, if not theatrical, moments.

Thursday, January 12 FBI Agent William Asbury takes the stand for the prosecution. He testifies about the various fossils, maps, documents, and other matter seized during searches of the institute. As he does so, the prosecution formally introduces many of these pieces into evidence. There are so many exhibits—more than 500—that the prosecution provides the jurors with notebooks cataloguing the evidence. Six television monitors also display institute field notes for jurors and spectators. Asbury testifies that these notes indicate that the defendants knew that they were illegally collecting on public lands.

Friday, January 13 Duffy cross-examines, or tries to cross-examine, Asbury about some of the fossils in question. Example: Did Asbury know that the institute had sold one of the specimens to the Smithsonian, charging only what it cost to prepare it because the fossil was of scientific import? Zuercher rises to object, arguing that how much the institute charged is irrelevant. Judge Battey sustains the objection. Still, Duffy has made his point to the jury; the Larsons weren't profiting from this so-called illegal activity. Why, therefore, the fuss?

Tuesday, January 17 During his cross-examination of Asbury, Bruce Ellison focuses on another fossil. The Larsons are charged with falsely telling the customs service that a baleen whale excavated in Peru had no commercial value. After bringing the bones back to Hill City, the institute reconstructed the fossil and then sold it for $225,000. Ellison asks Asbury if he knows how many man-hours went into the excavation, transport, and preparation of the specimen. His point is that the bones truly had no value when they passed through customs; thus, there was no false declaration.

Ellison also tackles the obstruction-of-justice charge against his client. Does Asbury know when Neal Larson first changed the dates on the boxes of fossils in question? Ellison is laying the groundwork for his defense: that although Neal did change the dates, he had remedied the situation by the time he turned the boxes over to the FBI on the day of the raid. Asbury says he is not aware when the dates were first changed.

Friday, January 20 One of the counts of the indictment charges that Peter Larson and Eddie Cole collected a *Paleonephrops browni* (fossil lobster) from federal lands at the Fort Peck Reservoir in Montana, despite the fact that Larson was aware that others had been prosecuted for illegally collecting fossils there.

The prosecution presents Becky Otto, an archaeologist with the U.S. Army Corps of Engineers. She testifies that in September 1988, *after* the lobster was collected, the corps had turned down a request by the institute to collect at the reservoir The prosecution believes this proves that the defendants knew a permit was required.

On cross-examination, Duffy attempts to establish his client's innocence by demonstrating that Larson and the government differ over what constitutes illegal collecting. Otto says her agency never gives permits to commercial collectors, but she admits that Corps of Engineers' regulations don't specifically use the terms "fossils" or "paleontological resources." She adds that "minerals" are included in the regulations and that fossils were intended to fall under that definition. Duffy follows up by asking if Otto is aware that the solicitor general of the U.S. Department of the Interior has ruled that fossils are not minerals. Mandel objects; Duffy's portrayal of governmental policy here is inaccurate, the prosecutor asserts.

Ron Banks, Cole's lawyer, also cross-examines Otto. The archaeologist has stated that corps rules ban disturbing any government property or natural resources. Does that mean a person could be arrested for picking a flower on federal land? asks Banks, who is trying to plant the idea that the regulations are hazily, if not foolishly, worded.

Otto: "The possibility is remote."

Tuesday, January 24 Two important witnesses are examined, Sharkey Williams and Lee Campbell. Larson is accused of removing fossil remains, including portions of three *Triceratops*, from Williams's ranch in the days preceding the discovery of Sue. Williams testifies that he owns some of his land, but he leases the land in question from the Cheyenne River Sioux tribe. He says that Larson could have found out the boundaries of the leased land by asking officials of the tribe. He adds that he and Larson never talked about the difference between the leased and unleased land. Larson has always maintained that he believed Williams owned the land and had the power to give permission to dig. Williams testifies that he did give Larson permission to dig but did not know that Larson was actually going to remove fossils. He says that Larson later sent him a check in payment for the fossils and for disturbing the land. Williams cashed the check. Given these facts, is Larson guilty of a crime? The jury will have to decide.

Campbell is a commercial collector with whom Larson has done business. He has been granted immunity in return for his testimony. Campbell admits that he and Peter Larson removed a *Triceratops* from BLM property near Lusk, Wyoming, in June 1987. Larson has maintained that he thought they were on private land. Did they know they were on federal land? asks the prosecution. Campbell responds that he told Larson there was a "high probability" that they were indeed within BLM boundaries.

On cross-examination, Duffy produces a transcript of Campbell's grand jury testimony concerning the location of the fossil. "I've searched my brain 50 times on that, and I just can't remember," he had testified. Duffy wants to know why Campbell is now contradicting that testimony. "How many more times did you rack your brain?" he asks.

Campbell: "It's been another 50 or so. A lot." He explains that other evidence refreshed his memory. What evidence? A map on which he had marked an "X" to designate where the *Triceratops* was found. The mark is on BLM land, just across the border from the private land. When did he show the map to Larson? Duffy asks. Campbell admits that he can't recall. Again, the jury will have to determine if Larson committed a crime here.

Tuesday, January 31 The government calls Marion Zenker. Because she works for the institute, Zenker is termed a "hostile witness." As a result, the rules of cross-examination apply. Zenker demonstrates that she is not kindly disposed to the prosecution. She prefaces one response by saying, "Prior to our being invaded by federal agents in 1992 . . . ."

Most telling is Zenker's description of the manner in which the Larsons operate. Zenker began working at the institute in 1989. At that time, she tells the jury, the bookkeeping and correspondence were a mess. Most of the Larson brothers' contracts were verbal, not written. She even found some notes inscribed on the back of matchbook covers. In short: "These were people who loved fossils and happened to have a business," says Zenker. Eventually, the jury will have to decide whether to accept this "absent-minded professor" defense.

Tuesday, February 7 Sue Hendrickson testifies. Because the prosecution wasn't certain when it would need her, Hendrickson has been in town for several days. Still not allowed to communicate with anyone at the institute or view the trial, she is in emotional turmoil.

In his opening statement, Duffy anticipated Hendrickson's appearance. He told the jury to remember that she had been given immunity in return for her testimony. This admonition was a not-so-subtle hint that witnesses put in this position may do whatever it takes to please the government and save themselves from prosecution—and that their credibility should be weighed accordingly.

Mandel suspects that Duffy will grill Hendrickson about her own alleged sins for which the government has forgiven her. To defuse such cross-examination, Mandel himself asks Hendrickson about her past. The prosecutor inquires if Hendrickson always declared fossils, amber, gold nuggets, pearls, and the like when she entered the United States and if she paid income tax on items that she sold. She responds that she always declared items she believed she was required to declare and that she paid taxes on income she believed was taxable.

Mandel moves to Hendrickson's knowledge of the defendants' conduct. He inquires about fossils collected in Wyoming with Lee Campbell and others in 1987.

Mandel:	There was no permission obtained to collect this specimen, was there?
Hendrickson:	Not that I remember.
Mandel:	Were you aware that this specimen was collected from BLM land?
Hendrickson:	I don't remember being aware of that, no.

A collecting trip to Montana with Larson and Eddie Cole in 1988 is the next subject of inquiry. Hendrickson acknowledges that the party slept at a campground located at Fort Peck Reservoir, federal land. She does not, however, recall, collecting the lobster fossil that the government alleges was illegally excavated.

Mandel has better luck when following the paper trail rather than the fossil trail. Peter Larson is accused of failing to declare $15,000 in cash that he took to Peru. Customs regulations require declaration of over $10,000 in cash or travelers checks. Bank records showed that Larson withdrew the $15,000 in two separate transactions of $8000 and $7000. Hendrickson accompanied Larson on the trip.

| Mandel: | Do you know why it was split into two separate transactions under $10,000 each? |
| Hendrickson: | I don't know why the two checks were written separately. Whether there was one calculation of what was needed, then a further calculation what was needed, I don't know. |

Hendrickson testifies that she and Larson split the $15,000 between them when they entered Peru. Why? "For safety's sake, you would each carry certain amounts of money. And as long as you are carrying less than $10,000 a piece, there's no need to make a declaration of it."

Mandel isn't satisfied. "Is it fair to say, then, that one reason was to avoid making a declaration?" he asks.

Hendrickson: "That's likely one reason, yes."

Duffy chooses not to cross-examine Hendrickson. "We'd set her up [to the jury] as their star witness, and I'd spent hours preparing for her," he later said, "but [the prosecution] didn't lay a glove on her."

Ellison asks only a few questions. He focuses on two fossil finds. First, he tries to establish that the defendants honestly believed they were on private land when they excavated the *Triceratops* from Sharkey Williams's property.

> Ellison: From your conversations with Mr. Williams, did you have the impression that that was his ranch . . . ?
>
> Hendrickson: Yes.

Hendrickson leaves the stand, deeply dismayed that she was put in the position of having to testify against her closest friends.

"I felt so sorry for her," Larson would later say. "I knew she didn't want to hurt us. And she really didn't. I think her testimony was a dud for the government. She knew we hadn't done anything wrong."

Thursday and Friday, February 16–17 Under cross-examination, defendant Bob Farrar, the Larsons' partner, concedes that the institute didn't get permits at many of the sites where it dug and that it did not always require those from whom it bought fossils to verify that their finds had been collected legally.

The government charges that the institute sold the Field Museum a catfish fossil purchased after it was collected by Fred Ferguson in Badlands National Park.

> Mandel: For example, when you are buying fossils from within the boundaries of Badlands National Park, sir, could you go so far as to, say, call up the park ranger and ask if it's legal?
>
> Farrar: No, I wouldn't bother him with that.
>
> Mandel: Because he might tell you that it isn't, right?
>
> Farrar: I don't know.

But Farrar did think that Ferguson had "some sort of permission . . . to collect."

> Mandel: You, of course, have seen that written permission, haven't you, Mr. Farrar?

Farrar:	No, I haven't.
Mandel:	Did you ever ask to see it?
Farrar:	Yes, I have.
Mandel:	You weren't shown it.
Farrar:	No.
Mandel:	Well, did that make you possibly question whether or not it existed?
Farrar:	No, I believed it did exist.

Is Farrar naive or was he calculating? This is a question for the jury.

Mandel offers another example of such naiveté or calculation—the institute's participation with Campbell in the removal of a *Triceratops* from what turned out to be BLM land. The institute has always insisted that Campbell never said the fossil was on public land.

Mandel:	Before collecting it, what effort did you make to find out whether or not that was federal land?
Farrar:	We made no efforts whatsoever.

The prosecutor then turns to the *Triceratops* skull found on Sharkey Williams's ranch. The institute eventually sold the fossil to a Japanese museum for $125,000. Before assenting to the purchase, the museum asked for certification that the skull had been collected on private land.

Mandel:	Tell me, what efforts did Black Hills Institute make to determine the status of the land that came off of?
Farrar:	We asked permission from the rancher to collect; that's all we did.
Mandel:	No one ever made a five-minute phone call to the tribal realty office to find out the status . . . ?
Farrar:	Not that I know of.
Mandel:	No one ever checked with the BIA [Bureau of Indian Affairs]?
Farrar:	Not that I know of.
Mandel:	No one ever checked with the Register of Deeds?
Farrar:	I don't think so.

| Mandel: | Well, tell me, would you buy a house for $125,000 without checking whose land it was sitting on, sir? |
| Farrar: | No, I would not. |

Mandel addresses other charges—the undervaluation of specimens shipped to Japan and obstruction of justice. Farrar testifies that he undervalued by half a *Triceratops* skull sold to a Japanese concern for $80,000. He acknowledges that U.S. regulations require listing the "sales price," but he explains that his shipping agent requested he do something different. "He wanted the shipping documents to reflect a lower-than-actual price," says Farrar. Why? "What I recall is that he wanted it shipped at what we would consider its replacement cost."

Given an opportunity to examine Farrar after Mandel finishes, Ellison returns to the excavation of the suspect fossils. Farrar states that it was his understanding that all lands involved were privately owned.

| Ellison: | Has it been the policy of the Black Hills Institute and your policy to accept the word of the landowner as to what constitutes his land? |
| Farrar: | Yes. |

Friday, Tuesday, and Wednesday, February 17, 21, and 22 Farrar's testimony has not advanced the cause of the defendants. ("Bob just fell apart," Peter Larson would say later. "He was admitting to things we hadn't done. He just wasn't properly prepared for the way in which the questions were asked.") This places an additional burden on the two men whom everyone expects will be the final defense witnesses, Neal and Peter Larson. Neal, who will be questioned by Ellison, takes the stand first, for questioning that will last three days. Jurors will later say that his testimony was critical to their decision-making process.

Neal has been preparing for this moment for days with his attorney. Still, he looks uncomfortable as he takes his seat and tries to adjust the witness microphone.

| Ellison: | Little nervous this morning? |
| Larson: | I am extremely nervous, sir. |

Because of the Fifth Amendment protection against self-incrimina-
tion, none of the defendants can be compelled to testify. Why, Ellison
asks, is Larson testifying? "Because for a number of years we have been
alleged to have committed crimes. I wanted to have the chance to tell my
side of the story and to tell the jury, to tell the people what happened."

Ellison's first goal is to present Neal as the honest, hard-working
everyman that most people believe him to be. Then Neal can tell "his
side of the story"—rebutting the government's case charge by charge.

The attorney begins by asking why Neal collects fossils.

Larson:	. . . because I love them. It is something wonderful to go out into the prairie or into the hills and to look for something that nobody else but God has ever seen before. . . .
Ellison:	Were you out thinking you would strike your fortune?
Larson:	To us, to my brother and myself . . . the pleasure and the wonder of hunting for the treasures, for the pleasure of what you can find in the earth, it's just overwhelming.

Neal explains that when he began at the institute out of college, he
earned $100 a month. He currently draws a salary of $2400 a month, or
$28,800 per year.

Ellison establishes that his client is a man to whom family values are
important. Neal testifies that he grew up in a large, loving, fossil-collecting family and is now the happily married father of five children
ranging in age from 4 months to 20 years old.

He is a religious man as well (as was shown by his desire to go to
church before bringing supplies immediately after Sue was found). Neal
explains to the jury that he went to a special high school so that he could
become a Lutheran school teacher. Ellison observes that such an occu-
pation is "pretty unrelated to geology." Neal respectfully disagrees,
"because all of geology and all of the things that we find on earth I
believe were created by God."

After Neal details the history of the institute and the nature of its
business, Ellison attempts to question him about the institute's "scien-

tific recognition." The attorney has more than a dozen notebooks demonstrating the institute's relationship with museums, donation of specimens, and authorship of scientific papers. Zuercher objects to the evidence. Judge Battey asks Ellison to explain: "How is that relevant to whether or not they broke the law?"

Ellison:	It's to rebut the suggestion in the indictment that the scientific aspect of the institute is simply for greed and that it's a ruse for them to go out and unlawfully collect fossils.

After a lengthy exchange Judge Battey tells Ellison, "You are trying to get the jury to decide this case for the wrong reason." Ellison, of course, disagrees. The judge tells him to ask a "few general questions; then we are through."

Ellison eventually moves to the manner in which the institute conducts its collecting. Larson testifies that as a boy he and his brothers collected on neighboring land.

Ellison:	When you were collecting on other ranchers' land, did your dad instill in you how you should treat the question of permission?
Larson:	You must always get permission whenever you go on anybody's property.
Ellison:	Is that something you've tried to do ever since you were a little boy?
Larson:	Yes, it is.

Ellison asks if there came "a time when the Black Hills Institute actually developed a policy of practice regarding collecting on federal land." Larson explains that in the mid-1980s his brother Peter worked closely with South Dakota's U.S. Senator Larry Pressler to develop legislation governing such collection. The effort was in response to BLM regulations that would have allowed commercial collectors on the land but made it very difficult for amateurs. Pressler was responsive to fashioning a compromise giving greater access to amateurs. He introduced a bill, but it didn't pass. The experience persuaded the Larsons that there was antagonism to commercial collectors. As a result, Larson says, "Our policy was to avoid collecting vertebrate fossils and then later of all fossils on any federal lands to avoid any confrontations as we seem to have got ourselves involved in now."

In 1988, the institute's board of directors had formally adopted a policy regarding collecting on federal lands. This action, Larson explains, followed the three-year effort of the National Academy of Sciences to "formulate an agreement between amateurs, professional commercial collectors, academics, and government agencies." Although never implemented, the NAS report had recommended that the federal agencies should allow collecting on public lands, with the exception of national parks and monuments. Amateurs were to be allowed to collect fossils from public lands and a permitting system was to be established to regulate professional commercial collectors.

Ellison: So was there a decision made by the Black Hills
                Institute to seek some of these permits and see if
                that's in fact what the agencies would do?
Larson: That's what we tried to do, yes.

And what happened? With one exception, "all [permit requests] were immediately denied," says Larson.

Ellison begins questioning Larson about each allegation of alleged illegal fossil collection. According to Larson, in all but one instance, the institute received permission to dig from the landowner. In some instances, the institute personnel relied on experienced collectors and skilled map readers with whom they had previously worked to determine that they were on private, not public, land. This was the case with Campbell, for example. In other instances, however, the Larson brothers themselves used their own maps to make sure they were on private land.

Some of these maps can be difficult to read. BLM maps are color-coded, Larson explains. On a BLM map Larson and Wentz used on a collecting expedition near Edgemont, South Dakota, national forest and national grasslands are designated in green. Patented lands—those owned by individuals and on the tax rolls—are white. State lands are blue. There is also land shaded in pink, a rarity on such a map; it is designated "Bankhead Jones," apparently a reference to previous owners. On the map, the words "Buffalo Gap National Grasslands" cross the areas marked by pink, white, and green.

Larson was interested in exploring a large exposure in the Bankhead Jones area. After conferring with a local rancher, Larson believed that the exposure lay on land owned by a grazing association. It turned out this was forest service land. Larson and Wentz eventually collected parts of a mosasaur and, in his words, other "junk," from the property.

Ellison:	Did you understand, sir, when you looked at the map that the pink was forest service land?
Larson:	No, sir.
Ellison:	If it had been forest service land, would you have gone on that land?
Larson:	No, sir.

Jurors listening to Farrar might have concluded that the institute made little effort to determine whether they were on tribal lands. To demonstrate that this was not the case, Ellison examines Larson extensively about remains of a *Triceratops* allegedly removed illegally from Standing Rock Reservation in Corson County, South Dakota. Larson details a relationship with members of the Standing Rock Sioux tribe that began in the 1970s when a rancher named Emma Bear Ribs gave him permission to collect ammonites on her property.

In 1983, Larson says, he received permission to collect from another Sioux rancher, Wesley Arnold. He was excavating some dinosaur bones when he was approached by a tribal officer, Paul Bruno Red Dog. Red Dog asked him whose property he was on. Larson said, "I'm on Wesley Arnold's property."

According to Larson, Red Dog responded: "Well, the land in this area is kind of checkerboarded and you cannot always be sure. Many times the ranchers don't know if it's their property or if it's tribal property."

Mr. Red Dog suggested that Larson cease collecting and get a permit from the tribe. Despite Arnold's continued assurances that he owned the land, Larson stopped collecting until the tribe awarded him a permit. For several years thereafter, he always called Red Dog or his successor, Ron Yellow, to tell him where and when he would be collecting.

The Standing Rock *Triceratops* that was the subject of the indictment came from badlands on the ranch of Tim Monnens. Monnens assured Larson that he owned the land, gave him permission to dig, and took money for the fossil, Larson says. This seems to contradict Monnens's testimony earlier in the trial. When called as a witness by the government, Monnens said he never told Larson that he owned the land in question. Ellison does not ask Larson if he checked with the tribe before dealing with Monnens.

After addressing each alleged fossil transaction, Ellison moves to the financial transaction. Neal acknowledges that he signed the two separate checks totaling $15,000 that his brother took to Peru. Why two checks? Neal explains that at first his brother thought $7000 would cover expenses. But, then: "[Peter] informed me that he had to take more money . . . that $7000 was not enough, because they were going to be there for an extended amount of time."

Ellison:	Why didn't you tear up the first check?
Larson:	We try not to void checks; with the additional check, it made the same amount . . .
Ellison:	Did you write these two checks to try and avoid any reporting?
Larson:	I did not.

Finally, Larson gets the opportunity to tell his side of the story regarding the alteration of dates on boxes of fossils. He acknowledges that he changed the dates on some boxes in anticipation of an FBI raid. The boxes had nothing to do with Sue. Instead, they contained bones removed from Sharkey Williams's ranch at approximately the same time Sue was excavated. The labeling on the boxes of fossils from Maurice Williams's ranch and Sharkey Williams's ranch was similar.

Larson testifies that prior to actually being served with the search warrant in May 1992, he had no understanding that the government was interested in any fossils besides Sue. Worried that the FBI might mistakenly take the Sharkey Williams's boxes, he changed the date.

| Ellison: | So the intent was not to prevent the government from what they thought they were coming for, but from getting the wrong thing? |
| Zuercher: | Objection, leading. |

Judge Battey sustains the objection, but Larson and Ellison have made their point.

Ellison now asks if, after seeing the search warrant, Larson tried to hide the boxes he had altered. No, says Larson. He directed the agent to the area where the boxes were stored and "commenced to show them the different items that were on the search warrant."

Larson has come across as decent man who loves what he does for a living. Ellison's gentle questioning seems to have established reasonable doubt as to whether the defendants knowingly and willfully collected fossils illegally. But can Larson hold up under Zuercher's sure-to-be-aggressive cross-examination?

The prosecutor revisits each of the acts recounted in the indictment. He begins at the Buffalo Gap National Grasslands, where Larson said he removed fossils from what he mistakenly thought was property owned by a grazing association.

Zuercher:	When did you become aware that the ranchers used the *national* grasslands for pasture?
Larson:	I don't think that question ever came up in my mind to come up with an answer on that, sir. . . .
Zuercher:	. . .Were you aware that you could not remove vertebrate fossils from forest service lands?
Larson:	No, sir.

An incredulous Zuercher shows Larson a copy of the institute's unsuccessful application to the U.S. Forest Service requesting a permit to dig fossils on the grasslands. Wouldn't the fact that the institute applied for a permit suggest that Larson knew it was illegal to dig without one? Larson says he is unfamiliar with the application and the particular regulations.

Zuercher: As an officer of the company, didn't you sign the
                board of directors' minutes indicating that
                indeed you wished to submit such an application
                for a permit?

Larson: Yes, I did.

Before leaving the grasslands, Zuercher turns to Larson's reading of
the color-coded map.

Zuercher: When you saw "Buffalo Gap National Grasslands"
                written on the map right over the top of it, didn't
                that give you some reason to think that perhaps it
                was Buffalo Gap National Grasslands?

Larson: It's also over the white, too, sir, and the white is,
                according to that, patented land.

Larson does admit that he never called the forest service or local
register of deeds to find out whose land he was on.

Zuercher moves to the lobster fossil allegedly removed from the
Fort Peck Reservoir. He notes that in 1983, Peter Larson had gone to
Washington, D.C., to protest the arrest of David Anderson, who had
been charged with illegally removing fossils from this same federal land.
Wouldn't this have put the institute on notice that fossil collecting there
without a permit was illegal?

Larson: Depending on what he had been cited for. He was
                cited for violating the Antiquities Act, which at
                that time had been found to not include fossils.

Zuercher also questions Larson about the *Triceratops* from the
Standing Rock Sioux Reservation.

Zuercher: You certainly don't disagree with Mr. Monnens
                when he indicated he never told you that was his
                land do you?

Larson: He never—he indicated that was his land. He told
                us that was his land. He never told us otherwise.

Larson says he paid Monnens $500 for the fossil.

Zuercher:	Did you ever tell the Standing Rock Sioux tribe that you sold this *Triceratops* for $80,000?
Larson:	No, I did not.

This is not relevant to whether the specimen had been illegally collected. Rather it is intended to suggest that the self-effacing scientists from the institute are in reality profiteers. The sale of other items, like the duck-bills from the Mason Quarry, is, says the prosecution, further proof that institute personnel talked their way onto land and took fossils under false pretenses; they gained access by playing up their scientific interest in exploration and then, after paying the landowner little or nothing, sold their finds for large sums of money.

The institute's response is that they always indicated they were commercial collectors, and, further, that thousands of hours of preparation were required to turn the *Triceratops* bones and other finds into costly museum specimens. This response does not trumpet the Larsons' pride in being capitalists. Outside observers might congratulate the brothers for making a profit in what Henry Fairfield Osborn called "a profession [that] is seldom if ever, remunerative." But the defense strategy here and throughout the trial is to present the brothers to the jury as self-made scientists, not self-made men.

Zuercher questions Larson about another *Triceratops*—the one removed from Sharkey Williams's ranch. Larson acknowledges that he never checked with the BIA or government to verify that the land was private. "The efforts had already been made," he explains. "We had asked for and received permission from Mr. Sharkey Williams. We believed it to be his property."

Zuercher shows Larson a map that shows that the land where the *Triceratops* was found was owned by the Cheyenne River Sioux (and leased by Williams). The map is from the institute. Larson says he never saw it. (To this day, the Larsons maintain that "this map was false evidence, a map we never owned." They add that "the government did not even give it to us with the evidence they returned" after the trial.)

Ellison conducts redirect examination after Zuercher finishes. Moments before Larson leaves the stand after three days of testimony, his attorney tries to clear up confusion about whether Marion Zenker checked with the Ziebach County Register of Deeds to determine the ownership of Maurice Williams's or Sharkey Williams's ranch. For the first time in the trial, Sue, the *T. rex*, is the subject of questioning. The exchange that follows seems a fitting climax to the proceedings.

Ellison:	That was the first complete *Tyrannosaurus rex* in the world, was it not?
Zuercher:	Objected to as irrelevant.
Judge Battey:	Sustained.

To the surprise of many in the courtroom—including Judge Battey, who said he had not finished compiling jury instructions—Duffy announced that Peter Larson would not testify. "I wanted to testify for the same reason as Neal—to tell my story," Larson later said. "And I expected to up until the last minute. Unfortunately, there was a problem."

That problem was a homemade video shot late one night in Peru. Larson says that a small portion of the tape, which had been seized by the government, featured offhand remarks about certain individuals. He acknowledges that those remarks were in questionable taste and might have offended the jury. He insists, however, that the tape had nothing to do with any of the charges against him.

The prosecution wanted to show the jury the tape. "Judge Battey refused but said he would revisit his decision if I ended up testifying," Larson explains. Confident that Neal's appearance had saved the day, and unsure if Judge Battey would keep the embarrassing piece of video out of evidence, Larson decided not to take the stand.

The following day, February 23, the government and all five defense lawyers presented their closing arguments. Each side had three hours to make its last appeal to the jury. Mandel used all of his time. Speaking without any notes, he reviewed each charge and the corresponding evidence that, he argued, demonstrated the defendants' guilt. He returned to the theme he had struck in his opening statement—that the defendants knew it was illegal to collect on public lands and that they knew they were on public lands when they collected most of the fossils. If they didn't know, he said, they should have known. "They had the ability to read maps, make records, and know where they were. . . . These guys were professional fossil hunters." He added that Judge Battey would instruct the jury that the defendants "can't turn a blind eye to everything and not look. They can't remain intentionally ignorant and avoid intent."

Mandel would later say, "I felt the evidence had been presented to convict the defendants on every count." Still, he knew he had an uphill battle convincing the jury. For almost three years Duffy had argued the case in the newspapers. "He spun it as big government picking on the little guy." But as modestly dressed and self-effacing as they appeared in court, the defendants were not "little guys," according to the prosecutor. They sold many fossils for hundreds of thousands of dollars. Yes, thousands of hours went into the preparation of these specimens, but most of the preparators employed at the institute worked for near the minimum wage, said Mandel. Even if it cost $5000 to $10,000 to prepare a specimen that sold for over $100,000, that was quite a markup. "It makes cocaine look sad," Mandel said later. (The Larsons dispute Mandel's figures. A duck-bill that sells for a maximum of $350,000 takes 15,000 hours of work and requires tens of thousands of dollars in glue and other preparation materials, they say.)

In his closing argument, Duffy echoed words uttered by Mayor Vitter over a year earlier. "Either Peter Larson is a criminal mastermind the likes of which this state has never seen or there has been a terrible, terrible travesty."

Ellison asked the jury to consider the depth of the evidence, not the weight. "They wanted to overwhelm you," he said, referring to the prosecution's lengthy parade of witnesses and exhibits. But despite this parade the government had failed to demonstrate a conspiracy, said Neal Larson's attorney. Mandel's argument notwithstanding, the prosecution had never shown that the defendants possessed criminal intent, he continued. "They're not crimes, they're mistakes." He concluded by noting that the institute was the best preparator of fossils in the world. "Only the government could take something so wonderful and make it sound so bad."

A criminal mastermind or a terrible travesty? Crimes or mistakes? Wonderful or bad? After the longest, costliest criminal trial in South

Dakota history, the fate of the Larsons, their fellow defendants, and the institute finally rested with a jury of their peers.

The defendants went home to Hill City to wait for the verdict. One week passed. In the middle of the second week, the jury informed the judge it was having difficulty reaching a decision on some of the charges. Keep trying, said Battey. The second week passed. "It was incredibly nerve-wracking," recalls Larson. The institute was in the middle of mounting a huge dinosaur display for a special exhibit in Japan for the Tokyo Broadcasting System, which, along with the Houston Museum of National History, was one of the few clients that had retained its services since the indictments. "We were trying to work, but as you can imagine we weren't very efficient," Larson says. "I couldn't sleep."

Larson had hoped that the jury would see that the government had no case and render a swift verdict for the defendants. Now he worried that the lengthy deliberations signaled that the jury was probably going to find him guilty. Mandel thought just the opposite; the longer the jury took, the less optimistic he became.

On March 14, in the middle of the third week, Duffy called Larson. "The jury's reached a verdict," he said. The Larsons, Farrar, Wentz, and their friends and families nervously hopped into their trucks and headed to Rapid City.

11

I KEPT WAITING FOR SOMETHING TO HAPPEN

"Five million one hundred thousand. Five million two hundred thousand."

And then another pause. This one even longer than the one at five million dollars.

Redden scanned the horizon for signs of life. The paddles had suddenly become as extinct as the *T. rex*.

Judge Battey had given the jury a sheet listing each count of the indictment and the particular defendants accused of the crimes alleged within each count. There were empty boxes by each defendant's name under each count. The jury was to fill in the appropriate boxes to indicate whether it had found a defendant guilty or not guilty or had been unable to reach a unanimous verdict. Foreperson Cecelia Green now handed the sheet to the bailiff, who presented it to the judge, who proceeded to read each and every one of the 154 separate verdicts. The defendants followed along, keeping score on their own verdict sheets.

Count 1: No guilty verdicts. Count 2: No guilty verdicts. Count 3 brought the first convictions: Peter Larson and the institute were found guilty of retaining (buying) fossils valued at less than $100 taken by a third party from Gallatin National Forest.

Although this was only a misdemeanor, Larson felt the air go out of him. He began to slump in his chair. "Keep smiling," Duffy whispered.

"This could be a lot worse." Larson took a deep breath, sat up, and allowed himself a thin grin.

By the time the judge was midway through his reading, the defendants had some reason to smile. The jury had not found any of them guilty of any felony with respect to their fossil-collecting activities—the activities that had triggered and fueled the government's three-year investigation. Indeed, only two additional misdemeanor convictions had been returned; Peter and Neal Larson were each found guilty of stealing fossils valued at less than $100 from the Buffalo Gap Grasslands—the land Neal said he had thought belonged to a grazing association.

Duffy was "euphoric." "The government spent millions and millions of dollars for a few misdemeanor convictions on the fossil counts," he would soon tell the press. He was right. The Larson brothers, Farrar, and Wentz were virtually vindicated of all charges that they were involved in a multistate conspiracy to steal, buy, or sell fossils from public lands. The jury had obviously failed to accept the government's argument that the institute was a criminal enterprise.

Judge Battey continued reading the verdicts, moving to the "white collar" counts related to customs violations, wire fraud, and money laundering. When all was said and done, Terry Wentz had not been convicted of a single crime. Neal Larson had been found guilty of the single fossil-collecting misdemeanor. The institute itself had been found guilty of four felonies—three customs violations and "retention" of the Badlands National Park catfish fossil it secured from the collector Fred Ferguson. Bob Farrar had been convicted of two felonies, each related to undervaluing fossils on customs declarations. And Peter Larson had been convicted of only two of the 33 felony charges he faced, each of them customs violations: failing to declare on a customs form $31,700 in traveler's checks he brought into the United States from Japan and failing to report $15,000 in cash taken to Peru.

The final tally: Of 154 charges, the defendants, including the institute, were convicted of eight felonies and five misdemeanors. The jury voted to acquit on 73 of the remaining charges and was unable to reach a verdict on the other 68 offenses. Jurors would later report that the vote was 11-1 in favor of the defendants on these unresolved charges; the lone holdout was foreperson Green. "Many of the jurors were crying as the verdict was read," Larson recalls. "It seemed clear they were unhappy with finding us guilty of anything."

Before leaving the courthouse, Duffy checked the sentencing guidelines for the customs violations convictions and concluded, "Peter was facing zero to six months in prison." Probation and perhaps a fine seemed the logical punishment. "No one ever goes to jail for this, I told myself," Duffy remembers. "It was a no-brainer. These were two incredibly Byzantine customs violations. If you'd have given me four counts (two felonies and two misdemeanors) to pick, I would have picked these four. They had no moral stigmata."

Despite Duffy's continued euphoria, Larson was shaken. Prison remained a possibility, as did the chance of a potentially business-crippling fine. As they left the federal building and the press approached, Larson told Duffy that he didn't think the institute could afford one guilty conviction. Many clients had been waiting for the verdict before deciding whether to continue doing business with the Larsons. How would they react now? And how would existing clients like those in Tokyo respond?

Duffy paused just outside the courthouse. "I've got something that will change this thing forever," he told his client. He pulled a long, thin cigar from his pocket and lit it. Then he moved towards the waiting reporters.

What do you think of the verdicts? he was asked.

"I'm so pleased I can't even begin to describe it," he said.

"Is that a victory cigar?"

"You got that right," said Duffy.

In the coming months, some who had followed the case would criticize Duffy for making such a bold gesture. After all, these critics noted, the client had been convicted of two felonies and still faced sentencing. Says Zuercher: "The last time I checked, when a client is convicted of a felony, that isn't a victory."

The prosecution—offended by the cigar—would play an important role in recommending a sentence. The judge—who it was clear had never appreciated Duffy's public proclamations—would have the final say. Why risk antagonizing these parties who held Larson's future in their hands?

Duffy has always defended the action and continues to do so. "I knew there would be resentment and jealousy and I'd be accused of histrionics," he said years after the trial. "But I'd do it again. I needed to leave them something to remember." He explains that the institute's financial future was in danger if existing and potential clients sensed the government had prevailed. Contracts were on the line, including the one with the Tokyo Broadcasting System. For months, it had seemed, said Duffy, that the institute was "in the indictment business instead of the fossil business." To keep the institute afloat, that negative image had to be erased. "Thank god for the tobacco leaf," he says.

Chicoine has a different take: "My memory is of a lawyer with a cigar calling attention to himself—challenging the government before sentencing in essence to prove that he didn't have a victory while his sheepish client is standing next to him with no control over the situation."

The prosecution also faced the media. A philosophical Mandel told reporters that he trusted the jury's verdict. "I felt we proved the case," he said in a later interview, "but if I was the only person I ever had to convince, we'd reach decisions very quickly." He says it is apparent that the jury "bought the big government versus the little guy spin."

The New York TImes called the verdict a "significant setback for federal prosecutors." A *Denver Post* headline declared: "No Meat in Fossil Verdicts." And back in Rapid City, the *Journal* proclaimed: "Verdict Victory for Fossil Hunters." But would that victory continue through the sentencing?

More than ten months passed before the Larsons found out if they were going to prison or would receive probation. The jury's inability to reach a verdict on almost half the crimes created most of the delay. Because of the deadlock, Judge Battey had to determine whether or not to acquit the defendants of the outstanding 68 charges. He eventually ruled against acquittal, saying, "a reasonable jury could have found the defendants guilty of all those charges they were unable to agree on."

This decision gave the prosecution the option to retry the case. It also moved juror LaNice Archer to speak out. "'Reasonable' to me was not spending millions of dollars on a glorified trespassing case," she told Bill Harlan.

Later, Archer and six fellow jurors held a news conference. They reiterated that they had voted 11–1 for acquittal on all the unresolved charges. Then they offered a startling revelation: if they had it to do all over again, they would now acquit the defendants of everything. They had voted to convict only because they interpreted the judge's instructions to require that they disregard key elements of the defense, explained Archer.

"They [the defendants] are not fraudulent people," added juror Cindy Fortin. "They're a good group of guys. Very good. The government did not prove they were guilty of anything, with intent." Juror Phyllis Parkhurst added: "I sat there through the whole trial, and I kept waiting for something to happen. It never came." Archer described Neal Larson as a man devoid of criminal intent. "This is the guy who invented Flubber. He's the absent-minded professor." As for some of the real professors who had testified, the government's expert witnesses, Archer said, "I started to sense early on a professional snobbery and jealousy in the scientific community."

In May, Peter Larson traveled to Tokyo in May to mount an ambitious dinosaur exhibition. Hendrickson helped him set up the display. This marked the first time that the pair had been able to talk face to face since Hendrickson had been called before the grand jury three years earlier and told that she could not communicate with any of the potential defendants. "There was a lot of hugging," Hendrickson remembers. "Our friendship was as strong as ever," Larson agrees.

Museum officials and the public alike lavished praise on the Tokyo exhibition. A paper Larson had written for the occasion, "*Tyrannosaurus rex—Cranial Kinesis*," was also well received. Here, based on his bone-by-bone, joint-by-joint study of Stan's skull, he detailed for the first time how the creature's head moved. "Pete knows as much about this as anyone in the world," says Bakker. "It was important work."

Larson returned to the United States upbeat. This mood continued through the summer, thanks to mostly good news from the courtroom. Despite Judge Battey's ruling, the government decided not to retry the case. In a prepared statement, U.S. Attorney Schreier said a new trial would not be a prudent use of government resources. She said the convictions in hand constituted an "adequate basis for the court to impose sentence in the case."

Said Attorney Ellison: "Hopefully such a reasonable attitude will continue through sentencing."

Sentencing never became an issue with one of the defendants. In August Judge Battey threw out Farrar's two felony convictions and the institute's felony convictions as well. The judge cited a recent ruling by the U.S. Supreme Court that suggested that these convictions would be overturned on appeal.

The news was equally good in the field. Larson had returned to the badlands, albeit cautiously. He no longer trusted collectors or landowners to tell him the status of a particular parcel of land. He checked everything with local registrars and used a GPS (global positioning system) to determine his exact whereabouts.

In August, Steve Sacrison found another *T. rex* in the same formation where Stan and Duffy had been found. A smiling Larson noted that excavation would begin five years to the day after Sue had been found. "It's a wonderful thing to happen to us at this time," he told the press. "*T. rex* has been our icon for quite a long time through ups and downs, and all the downs are easier to take if you have ups like this." He added that the find "symbolizes our joy in getting back to work again and doing the things we love." He named the dinosaur "Steven" after its discoverer and announced that this *T. rex.* would stand by Stan in the new Black Hills Museum.

Two weeks into the excavation, Larson developed another snapshot for his dinosaur photo album. It's possible that the tyrant lizard was a cannibal, he proffered. Spines on Steven's dorsal vertebrae had been sheared off in a manner suggesting that they had been eaten by one of their own. The sites of the bites were similar to those on other animals that had been preyed upon; Larson called them the "T-bone and tenderloin" sections. The huge size of the "portions" also pointed to only one possible aggressor—a fellow *T. rex.*

Thanksgiving and Christmas 1995 came and went without Judge Battey passing sentence. So, too, did New Year's Day 1996 and January 10, the first anniversary of the start of the trial. On January 22, Neal Larson finally appeared in court. He faced up to 12 months in jail and a $1000 fine for his misdemeanor conviction. The government had argued for a stiff sentence because, it said, Larson engaged in a pattern of criminal behavior.

A courtroom full of supporters, including Larson's wife and five children, listened as Judge Battey declared that it was a "close call" whether to put Larson on probation or send him to prison. His decision: two years probation and the maximum fine of $1000. A tearful Larson hugged friends and relatives after the decision and told reporters: "I'm happy, joyful, and thankful. It's been a long four years."

When Peter Larson headed to court for sentencing on January 29, he knew he could face up to 33 months in prison. Federal sentencing is not easy to diagram. In determining the length of a sentence, a judge must determine which "level" to assign to each conviction. Each level offers a range of prison time. For example, level 6—the level Duffy thought applicable—is zero to six months. Once the judge selects a level, he or she has flexibility to choose a sentence within the level's range.

In November, Judge Battey, whom the defense had attempted to recuse four times, had held a hearing to help him determine Larson's sentencing level. Duffy argued for level 6 and probation. The government, whom Duffy had criticized at every turn, aggressively pushed for a much higher level and a sentence that included significant prison time, 33 to 41 months.

A federal judge can take several factors into consideration when determining the sentencing level. If prosecutors can show that there were aggravating circumstances in the case, the level can be increased. In this case, Zuercher alleged violations of Peruvian statutes and offered evidence not presented at trial. Duffy believed the presentation specious and called his own witnesses from Peru to rebut it.

Sentencing guidelines also permitted moving to a higher level "if the defendant was an organizer or leader of a criminal activity that involved five or more participants and was otherwise extensive." Here Zuercher continued to press the point that the jury had refused to accept: The institute was a criminal enterprise. The prosecutor named Neal Larson, Farrar, Zenker, and Hendrickson as ringleader Larson's "participants."

Larson and others were flabbergasted to learn that a decision by the U.S. Supreme Court permits federal judges to use acquittals *against* a defendant when determining a sentencing level. That decision allows the judge to weigh evidence from the trial using a different standard than the jury uses. A jury can convict only if it finds "beyond a reasonable doubt." A judge can take into account whether a "preponderance" of the evidence would have supported a guilty charge. This is a considerably easier standard to prove. "I still don't understand how a judge can find you guilty of the same charge a jury finds you not guilty of," says Larson. "How is that fair?"

Fair or not, Judge Battey had, in effect, done just that in setting the sentencing level. His choice: not level 6, but, depending on different computations, 17 or 18. As a result, Larson faced 24 to 33 months in prison if the judge did not place him on probation. This put Larson at a level higher than those for aggravated assault, involuntary manslaughter, and child pornography, complained Duffy. Resorting to another popular culture reference, Duffy said, "[Larson] has to be beyond Snidely Whiplash for this to happen."

Earlier in the winter, Larson had broken his leg falling down an icy stairway. On sentencing day, with his wife and children at his side, he hobbled into the crowded courtroom on crutches. Once seated, he heard Mandel argue for a prison term of 30 months and a fine of $50,000. Then the judge asked him if he wanted to say anything. He rose slowly from his chair and said, "I absolutely and positively accept responsibility for all of my actions. . . . [But] I never intended to do anything wrong."

Battey, who had received 190 letters supporting Larson, was ready to rule. He said Larson was "a likable person," but he did not believe this protestation of lack of intent. Larson was too expert a collector not to have at least suspected he was on public lands, said the judge. In addition, Battey saw a pattern of conduct. "This was not an isolated theft of government property," he noted, in direct contrast to the jury's findings.

Judge Battey sentenced Larson to two years in prison. Coincidentally, this was the exact amount of time the government had offered in its first plea bargain offer. The judge also sentenced Larson to an additional two years of "supervised release," the equivalent of parole, and fined him $5000.

Duffy lit no victory cigar this time. "It's a staggeringly disproportionate sentence for the offenses committed," he said, promising to appeal the convictions. Later Duffy would take some solace in the fact that the fine had been relatively light. The judge could have put the institute out of business if he had chosen to, the lawyer said.

Larson was left to wonder how the plea bargain that would have kept him out of jail went up in smoke. He feels that the *Journal*'s story about the negotiations doomed the deal and still wants to know who leaked the story to reporter O'Gara. There has been speculation that someone from the defense camp did so, perhaps without thinking of the consequences or perhaps out of a desire for publicity. Duffy denies any involvement.

Larson recalls his attorney confirming details of the negotiations to O'Gara. This suggests that O'Gara might have learned these details from the government. But why would the prosecutors have leaked the story? Perhaps, some say, because there was a split in the U.S. attorney's office over the plea bargain. This theory holds that those who wanted to go to trial and put Larson behind bars called O'Gara in an attempt to sabotage the negotiations. If so, it appeared to have worked. "The bottom line is that whoever did it cost me two years of my life," says Larson.

Larson's stiff sentence caused many institute supporters to second-guess Duffy. In particular they questioned his decision from day one to play the case out in the press—a strategy that clearly antagonized both the prosecution and Judge Battey. Aware of his critics, Duffy adamantly defends his strategy. "I sleep the sleep of the just," he says, adding: "We were invited to a press war by Schieffer and the DOJ [Department of Justice]."

Clearly. But did he have to accept the invitation? Yes, he says. "The net effect was positive." There were only a few convictions, and "the fines could have been so much worse. [Judge Battey] could have applied economic sanctions that destroyed the institute."

But what of the prison term? "The sentencing was draconian," says Duffy. "The use of alleged illegal violations in Peru to form a basis for relevance was wrong."

Duffy concludes: "I'm offended that if you piss the government off you have to pay. That's what started the case and that's what ended it."

Perhaps. But if a lawyer knows that the government will react negatively, even punitively, if "pissed off," shouldn't he try to avoid waging war in the press?

In the end, Duffy seems to have won the battle but lost the war. His strategy to portray the defendants as the victims of an evil prosecutor and vindictive government worked brilliantly with the jury. But it failed just as spectacularly with the judge and the U.S. attorney's office. This is not to say that if Duffy had never talked to the press, he could have kept his client out of prison. Few members of the community think two years in prison is a fair sentence for failing to declare cash or travelers' checks. But from the moment the FBI raided the institute—perhaps even before, when Rangers Santucci and Robins began their investigations— the government seemed bent on making an example of the Larsons, and a slap on the wrist wouldn't do.

Why did the government need to make an example of anyone? "The government was never able to create a uniform system of laws to cover [fossil collecting]," Duffy explains. "And what government can't get by directive, oftentimes it takes by indictment." Chicoine agrees, but he places some of the blame on Duffy's shoulders. "Mandel was an up-front guy. In my opinion, you could have eventually sat down with him and made a lot better deal than two years. The crimes alleged here were not inherently egregious—not child abuse, molestation, or bilking people out of savings." The spotlight should have been turned off so that "the government could settle without a lot of people looking at them. The government doesn't want people to think it's soft."

Kevin Schieffer was never accused of being soft. Thus his comments on whether he felt vindicated by the sentencing might surprise some of his critics. "There's not a whole lot of vindication when the law is stupid," he said recently. "I empathize with the institute in that that area of the law was screwed up. To their credit they lobbied to get it changed to what they thought would be better. But even if getting permits is screwed up, you don't ignore them. As a prosecutor . . . you can't walk away when you have guys as sophisticated as these . . . flagrantly ignoring the law."

On February 19, three weeks after the sentencing, friends and family of Peter Larson gathered at the Alpine Inn restaurant on Main Street for a going-away party. Larson had agreed to "surrender" at the federal prison in Florence, Colorado, two days later. "There was some laughing and some crying," he remembers.

He admits that he had cried considerably more than he had laughed in the preceding days. "At the point of sentencing, it became more real. I prepared myself mentally for prison. I closed myself in. I could no longer engage my family."

On February 22, Larson and his wife made the ten-hour drive from Hill City to Florence, which lies 110 miles southwest of Denver. They parked outside the gate of the wall-less minimum security prison, where about 80 percent of the 400 inmates were serving time for drug-related crimes. Larson and Kristin talked, held hands, cried, and kissed. Then Larson climbed out of the car and headed inside.

Task number one was to fill out a prison induction form. "Under 'reason for incarceration,' the guard put 'failure to fill out forms,'" Larson laughs. "That was what my crime was called. That's what I got 24 months for after all the accusations and all the horrible things they accused us of."

After filling out the form, Larson was given a number, a uniform, a bunk in a 32-man dormitory, and a work assignment—shoveling snow (despite the fact that he was still recovering from his broken leg and was walking on a cane). He couldn't shovel, but he had to stand outside while the shoveling was done.

Through past travails, Larson had learned "my key to survival is to work my ass off." Once situated in the prison, he started writing a book and several papers about the science of the *T. rex*. He also started plotting a way to get Sue back.

Any sale, or, for that matter, any other plan for Sue had to be approved by the *T. rex*'s trustee, the federal government. The Department of the Interior had entrusted that job to the Aberdeen office of the Bureau of Indian Affairs. In a newspaper report published before Larson went to prison, the BIA's Carson Mundy said that he had seen six unsolicited offers for the dinosaur. Five proposals came from companies that wanted to help Williams finish cleaning and preparing the bones and then assist him with the sale. Sotheby's, the prominent auction house, had also offered to sell Sue on consignment. Mundy said that none of the proposals were from museums but that all came from within the United States.

Was a sale within the United States required? No, said Mundy. As trustee, the government's role was to determine the "highest and best use" for Williams's benefit, not the benefit of the United States. If a higher offer came from outside the country—from either a museum or a private collector—the government could not stand in the way of a sale.

Kevin Schieffer, firmly ensconced in the private sector now, took note of the developments. "My angst and anguish of a professional nature had occurred years before," he says. "At this point, I was beyond that. I just thought, What will be will be."

Peter Larson, firmly ensconced in prison, was unwilling to give up the fight. "I had some spies on the outside telling me what was going on," he says. "I still thought there might be a way to get her back."

12

EVERYTHING CHANGED
THAT DAY

"Five million two hundred thousand," Redden repeated.

Dede Brooks signaled to the auctioneer. He smiled. Her mystery bidder was finally in the game.

"Five million three hundred thousand," Redden said.

On a warm June day in 1997, three men entered a warehouse in East Harlem to examine merchandise said to have a street value of at least $1 million. John McCarter, Jr., John Flynn, and Bill Simpson had come all the way from Chicago to determine whether the stories were true: that the goods—laid out on shelves and in boxes—were first rate, one of a kind, must have. They needed to check the quality of the stuff in person before finding someone to help them acquire it. They held it in their hands. They took photos. They took measurements. And then they left.

If they did go after the merchandise, McCarter would be in charge of the operation, including finding the silent partners to assist in the purchase. Flynn and Simpson were his experts, the ones who could prepare the goods so that they might become infinitely more valuable. "Is this something you want?" McCarter asked the two men as they stood in the parking lot outside the warehouse.

Each said yes.

And so the Field Museum began its effort to acquire Sue.

The Harlem warehouse belonged to Sotheby's New York. Maurice Williams had put the *T. rex* up for auction. Liberated after almost five years in the machine shop at the South Dakota School of Mines, Sue was scheduled to go on the block in about four months, on October 4. In the meantime, interested parties could inspect her by appointment only.

Williams's decision to sell Sue had not come easily. Although deluged with purchase offers, he had hoped to retain ownership of the fossil. "We planned on having it here," he said in a telephone interview after the auction. "I had a relationship with the people at the School of Mines. I'm disappointed." He explained that he had wanted to charge people money to watch the cleaning and preparation of the specimen, sell casts of the skeleton, and take Sue on tour. In detailing these plans, Williams made no effort to hide the fact that this was, as he had told *Primetime Live*'s Sylvia Chase, all about money. (He even tried, unsuccessfully, to charge money for the interview excerpted in this book. "What do you expect us poor people to do?" he asked.)

Williams says that a lack of money forced him to abandon these plans. "The fossil was in storage at the School of Mines. The government refused to pay storage costs, and I couldn't afford to," he said. "I asked, 'Is it gonna be dumped in the street?'"

Enter David Redden, Sotheby's executive vice president. Redden had been following the custody battle for Sue for some time. "I was fascinated by it," he admits. "It was a wonderful snapshot of many aspects of our social compact, wasn't it? The rights of paleontologists, Native Americans, the rights of the public—each of which can be debated." When it appeared that the Native American would emerge victorious, Redden contacted him.

The suave Englishman had never sold a dinosaur, but he firmly believed that Sue was, in his words, "an absolute candidate for the auction process." He explains: "I was intrigued by the challenge facing Maurice Williams and the federal government. It [Sue] was not something he was going to keep, I didn't think. It required administration. But how to move it to the next owner was a real challenge."

In 1995 and 1996, Redden talked with Williams on the phone and visited him in South Dakota to make the pitch that the auction process could best meet that challenge. He found Williams to be both intelligent and savvy. When asked if he needed information about Sotheby's, the rancher said, no, he was familiar with the auction house.

What did Redden tell Williams? "That the process could do three things: one, assign a value; two, sort out the various possible contenders; and three be a mechanism for bringing to the forefront candidates who might not be so obvious." These candidates, Redden acknowledged, might include private collectors rather than museums. Williams didn't care. If he was going to sell the dinosaur he says, he was going to sell it to the highest bidder, be that an institution or an individual.

Redden eventually sold Williams on selling Sue. In consigning the dinosaur to the auction house, the rancher turned down numerous offers from a variety of parties, including private collectors, museums, and corporations. One Las Vegas casino was reportedly interested in making Sue a featured attraction. In November 1996, Sue was released from her holding cell in Rapid City and sent east to her halfway house in Harlem.

If Sue seemed a natural to be sold by Sotheby's, she also seemed a natural to be acquired by the Field Museum. The institution first opened its doors in 1893. Known then as the Columbian Museum of Chicago, its stated purpose was the "accumulation and dissemination of knowledge, and the preservation and exhibition of objects illustrating art, archaeology, science and history." In 1905, the museum's name was changed to Field Museum of Natural History to honor its first major benefactor, Marshall Field (of the department store family), and to better reflect its focus on the natural sciences. Thanks largely to expeditions by its own scientists, the museum's natural history collections are considered among the best in the world, ranking the institution in the same category as the Smithsonian, the American Museum of Natural History, and the British Museum.

The Field has more than 20 million zoological, botanical, anthropological, and geological specimens. These include dinosaurs found by paleontologist Elmer Riggs during the golden age of dinosaur hunting. Just five years after opening its doors, the museum had hired Riggs, a confederate of Barnum Brown, to find a brontosaurus. Museums in New York and Pittsburgh were already drawing large crowds with their 75-foot-long skeletons of this ancient herbivore.

Riggs unearthed a big dinosaur in 1900, but it wasn't a brontosaurus, and it wasn't 75 feet long. It was a 90-foot-long brachiosaurus ("arm lizard"). Until recently, this plant eater was thought to be the largest land animal ever.

Unfortunately, Riggs's brachiosaurus was less than 25 percent complete. Insufficient for display, the dinosaur rested in storage for more than 90 years—until the museum's one hundredth birthday. Then, to celebrate the occasion, the Field unveiled a glorious skeleton created by supplementing the Riggs bones with sculpted material from a more complete specimen in the Humboldt Museum in Berlin.

Brachiosaurus was not the only Riggs find standing on the museum floor. In 1901, the paleontologist did unearth a brontosaurus (now reclassified as apatosaurus). This specimen was complete enough to display and has long been the main attraction of the Field's "Life Over Time" exhibit.

Why with two stunning dinosaurs already roaming its landscape would the museum be interested in Sue? Bakker provides the short answer: "They didn't have a *T. rex.*" Not a real one, anyway. In the late 1920s, the prominent paleontological artist Charles Knight created a series of colorful, detailed murals for "Life Over Time." The most famous of these murals depicts a battle between a *T. rex* and a *Triceratops.* Horner notes that Knight's depiction has flaws but was a valiant effort for its day. "Today, we'd make the legs more birdlike and the snout longer, and we might make the body lean forward more, and raise the tail farther," he writes in *The Complete T. Rex.* "But at least Knight got *T. rex* off its haunches."

Word that a three-dimensional *T. rex* would be auctioned off had gotten John McCarter, the Field's president and chief executive officer since October 1996, off his haunches. He is a friendly, balding man in his early sixties. A graduate of Princeton and the Harvard Business School, he earned his living as a management consultant in Chicago before coming to the museum. He was no stranger to the not-for-profit sector, however. From 1989 to 1996, he had served as chairman of the board of WTTW, Chicago's public television station.

Like Peter Larson, McCarter envisioned Sue as a centerpiece that would draw visitors to his museum's door for generations to come. "You need a big, defining exhibit," he says. "Do you know what the most visited object is in a museum? The Hope Diamond at the Smithsonian. It brings 'em in."

Bringing 'em in, says McCarter, has many advantages, not the least of which is generating revenue to support less glamorous endeavors. "The reason we do dinosaurs is so that we can do fish," he says. He explains that the fish in the museum's collection may shed more light on evolution than Sue or any other dinosaur will shed, but having a dinosaur draw like Sue would "allow us to have financial robustness and strength to fulfill the museum's mission."

Because Sue's bones had been in storage for over five years, McCarter wanted Flynn, the Field's head of geology and an accomplished dinosaur hunter in his own right, and Simpson, the museum's chief preparator, to inspect the fossil firsthand. "We wanted to see how it would stand up to preparation," says Simpson. If the bones were too soft and crumbly, too "punky," as the preparator puts it, the museum would not be able to clean them, study them, and reconstruct them for display.

Seeing Sue was bittersweet, Flynn recalls. "It was depressing to see such a specimen in an auction warehouse." At the same time, "there was excitement, especially seeing the skull. It was spectacular. In paleontology my joy comes from fossil discovery or seeing a beautiful specimen. . . . This was the Hope Diamond of fossils." In the parking lot outside the warehouse, Simpson told McCarter that he was "thrilled to see that the bone was hard. It was well preserved. [Preparation] could be done."

Flynn agreed. "We do not take any specimen into our collection unless it has scientific value and meets [some or all of] our goals of research, exhibition, and education. . . . Here, with the quality of preservation, the completeness of skeleton, you could ask questions you couldn't of any other *T. rex*. . . . It met all three goals."

McCarter had already discussed Sue with Judy Block, the chairperson of the museum's board of trustees. After returning to Chicago from New York, he told her, "This is serious." He quickly met with Block and the trustees' executive committee, who gave him the green light to, as McCarter says, "figure out a way to get it."

The first thing McCarter had to figure out was how much money he might need and who could provide the money. "Getting it didn't mean just having it in plaster jackets in the building," says Flynn. "We would clean it, mount it, develop an exhibition, and [conduct] world-class scientific research." All these activities would require funds above and beyond the purchase price.

The Field was already in the process of preparing an ambitious exhibition for 1999, "Underground Adventure." Here, visitors would take a simulated journey underneath the Illinois prairie. "Shrunk" to the size of an insect, the travelers would meet animatronic ants and spiders along the way. The cost of the exhibit would be $10 million. The funding source? Corporations, including Monsanto, which alone had donated $4 million.

McCarter knew he would also need corporate support to acquire Sue and prepare her for exhibition. One company immediately came to mind. McCarter sat on the board of directors of W. W. Grainger Corporation. So, too, did Fred Turner, then the chairman of McDonald's. At a Grainger board meeting, McCarter pulled Turner aside, told him about Sue, and asked whether the fast food giant might consider helping the museum acquire the giant carnivore.

Why McDonald's? The company was based in the Chicago area and already had a long-standing relationship with the museum. Equally important, McDonald's had a rich history of community involvement, particularly in supporting programs with an impact on young people.

McCarter remembers that Turner responded enthusiastically to his initial overture. "I didn't even have to give him my full spiel," says the museum director.

Back at McDonald's headquarters in Oakbrook, a suburb west of Chicago, Turner turned the matter over to a vice president, Jack Daly. Daly, a genial man in his early fifties, is not in charge of marketing but rather worldwide communications. "I'm involved in managing the reputation of the brand and the reputation of the corporation," he explains.

When Daly told his colleague Cathy Nemeth that he was going to see a man about a dinosaur, each laughed. "I basically thought it would be a courtesy call," remembers Nemeth, the home office's director of communications.

Daly confesses that he was not terribly enthusiastic as he made the hour long drive into the city. "I have an abiding interest in anthropology and paleontology, and I have three small kids," he says "I know what a significant cultural trend dinosaurs are. But I must say when I went down there I didn't think we'd find a commonality. . . . We're not the kind of company interested in playing a passive role. We're not just interested in a plaque. At the level [of financial commitment] we were talking about here, getting involved makes sense for us only if we can engage our customers."

McCarter knew this. As a result he was able to present Daly with several ways that McDonald's could engage the millions of Americans (and others around the world) whom the company serves. Casts of the dinosaur could be made and taken on tour, said the museum director. Knowing that McDonald's regularly created educational materials for schools, McCarter noted that Sue was a perfect subject for videos, books, and lesson plans.

Daly began to see some possibilities. "I liked the idea that this was big," he says. But many questions remained. How much money did McCarter think it would take to win Sue? What other companies might be involved?

McCarter told Daly that he wasn't sure how much Sue would command at auction—perhaps $1 million or more. He also said that McDonald's was the first corporation he had called.

"What about Disney?" Daly asked. He reminded McCarter that McDonald's and Disney had just entered into a ten-year marketing alliance. McCarter knew this, but he was not aware of one fortuitous element of that union: McDonald's was to be the sponsor of DinoLand USA, an integral part of Disney's Animal Kingdom, an adventure park scheduled to open in Orlando, Florida, in the fall. "John was intrigued by this," remembers Daly.

Daly had another question: What was the time frame for the preparation of Sue?

McCarter said that the unveiling might not take place until the year 2000. Daly saw this as bad news and good news. Two and half years was a lot of time to wait for a payoff on an investment. On the other hand, a millennium unveiling presented its own opportunities.

Daly had already spent a couple of years thinking about what McDonald's could do for the millennium. "It's kind of a hard thing to consider," he admits. "When you are a brand as ubiquitous and omnipotent as we are, you pretty much know there's going to be an expectation—that people are going to be asking: 'What is McDonald's going to be doing?' We were thinking, What could we do that would be impactful, be global, make the system proud?"

Building a children's hospital or the wing of a hospital was a possibility. And McDonald's and Disney had plans to bring 2000 children together to "celebrate what's good about the world and what could be better," says Daly. But he had been unable to come up with the "impact-ful" millennium project he desired. Could Sue be that project?

Says Nemeth: "Jack came back saying it may work. When he gets excited it's, well, he's the boss. It's contagious. We're a creative group by nature; nothing is unthinkable or unacceptable. Sometimes I think the more over the top it appears, the more we get excited by it. Our charge here is doing something different and unique and exciting. We have a saying here: 'Only McDonald's can do.' That may be overused and a bit cliché-ish, but it's certainly a benchmark of what you're look-ing for."

The more Daly thought about the project, the more it seemed like something McDonald's could do. "If John [McCarter] had said it might take ten years or even five, it might have been different, because my initial thinking was more as a millennium gift than anything else and everything flowed from there," he says. "I had been struck by something Bill Clinton had said about the millennium: 'If you are interested in celebrating the future, try to honor the past.' I liked the idea that Sue was this timeless, priceless thing and that if the museum could get it, it would last forever."

Daly called together Nemeth and about a half dozen others who made up his brain trust. "We started to break it into its component parts," he remembers. "We were thinking, Okay, so there's gonna be this big period of time—two years—when nothing happens. What can McDonald's do during that period?" The answer: Produce educational materials for schools and sponsor the McDonald's Fossil Preparation Laboratory on the museum floor where visitors could watch bone cleaning and preparation.

One imagines plans for Happy Meals and Sue action figures as well, but Daly and Nemeth insist that those options weren't seriously consid-ered. Such commercialization of a scientific specimen would, of course, need the museum's approval. But more than that, this wasn't primarily about marketing, says Daly, it was about *brand.* "The headline is: Give back to the communities you serve. It gets to the whole question of what's the full spectrum of a brand. The marketing piece is there, obvi-ously, because that's part of what a brand is, but there are a lot of other colors in the spectrum that add to it. Clearly, working in a socially responsible way with organizations like the Field Museum is an impor-tant part of the spectrum."

What kind of projects enhance the brand? "A perfect program for us is one with potential global impact in terms of news, in terms of goodwill, and then clearly one that has implications in the United States—one that can be worked right down into the grassroots of our organization—every town, every community," says Daly. Sue came close to fitting the profile.

One additional element could make saying yes to the museum even more appealing. Soon after meeting with McCarter, Daly traveled with Nemeth to Orlando for a previously scheduled meeting with their partners at Disney's Animal Kingdom. Sue was not on the agenda, but during a break Daly and Nemeth cornered Bob Lamb, the Disney vice president in charge of the new park. "Want to buy a dinosaur?" Daly asked.

Lamb, it turned out, was not only familiar with Sue, he had contemplated trying to acquire her. A few months earlier, a Disney designer had sent him a clipping about the *T. rex.* "I thought, This is a cool thing," remembers Lamb, whose experience with dinosaurs dates back to the opening of Epcot Center in 1982. As project coordinator for an energy exhibit that featured a dinosaur diorama, he had helped sculpt some of the creatures—"the side the audience couldn't see," he laughs.

Pursuing Sue was discussed at the highest levels, says Lamb. Eventually, however, Disney decided against bidding for her. "We didn't have the Field Museum connection or a connection with any other scientific institution," he explains. "And without that connection, this was not a good thing for Disney." Outbidding a museum for Sue and merely displaying her in DinoLand USA or any other venue would have been a public relations nightmare.

Now McDonald's was offering a scientific connection. Says Daly, "I'd like to say the process is totally systematic and strategic. I don't think it is. I think there's serendipity, and I think in this particular instance there was a lot of serendipity."

Within a few weeks, Lamb called to say that Disney would participate. They would feature a cast of Sue in DinoLand. And before the cast was ready, they would open a "branch office" of the McDonald's Fossil Preparation Lab in the park where museum preparators could work on Sue in front of visitors.

When Lamb called, McDonald's was still considering whether or not to get involved in the project. "The next thing you know they're calling to say, 'We're in,' which was like, Uh-oh, they're in, I guess it means we're in," remembers Daly. "It had a strange momentum. Any one of the players could have spun out, and that might have spun out the whole thing. But everybody sort of spun in."

Before spinning in, Daly did need, in Nemeth's words, "the buy-in of several key people in the company." Key among these were, she says, "our partners in the marketing department, who were very much involved with the Disney alliance."

Daly and Nemeth also had to persuade themselves that local operators would "buy in" and support the tour of Sue. These operators would end up footing the bill for Sue's visit to their respective regions of the country. "Our biggest concern was the receptivity of operators [who would ask]: 'How's this gonna sell hamburgers for us?'" says Nemeth, an attractive, energetic woman in her mid-forties. "We fight that all the time when we present a program that might have a cost to it that is really much more in the area of what we want to do for our customers in the community—brand building."

She adds: "This was not a program to sell hamburgers. . . . What we do for a living is make sure that we do help sell hamburgers or we probably won't have spots in the company rosters. But we also make sure we have a nice balance of programs that are meant simply to be effective to our customers in our communities and programs that are meant to drive traffic into our restaurants, and the truth is you need to do both and build credibility with customers."

Having secured the support of all necessary departments, Daly and Nemeth arranged a meeting with the person who would have final say on the matter, Jack Greenberg, the president of McDonald's U.S.A. (who has since become the corporation's chief executive officer). "We had a whole presentation ready," says Nemeth.

"You're here to talk about the dinosaur, right?" said Greenberg.

"Yes," said Daly.

"Is it a good idea, Jack?"

"Yes.

"Then okay, we'll do it," said Greenberg.

McDonald's and Disney were aboard, each with ambitious plans that would, in McCarter's words, give Sue "legs and reach." Now all the museum director had to do was get the dinosaur.

Despite significant financial commitments from both corporations, McCarter still wasn't certain if he had enough money to win Sue. Indeed, he still had no idea what she might bring at Sotheby's. He had been to only two auctions in his life—for cattle, not fossils. In each instance he had been a spectator, not a bidder.

Unwilling to be caught with his paddle down, McCarter picked up the phone and called an old friend from the WTTW board—Chicago art dealer Richard Gray. "Are you busy?" he asked.

Gray, a peripatetic community booster, wasn't surprised to hear from the museum director. "John's the kind of guy whose ideas won't stay still," he says.

Within the hour McCarter was at Gray's gallery, briefing the art dealer about Sue. "John needed some hand holding. He didn't want to get in over his head," remembers Gray, an old pro at the auction business, who once made a winning bid of more than $20 million for a painting on behalf of a client.

Gray quickly offered a strategy that would prevent McCarter from getting in over his head. "Try and buy up front," he said. "See if you can avoid the risk of auction. Talk to Williams. Talk to Sotheby's. Find out if you can buy directly."

McCarter's eyes lit up. "Can you do that? Can you preempt an auction?"

"It's not common, but sometimes it can be done. I've done it. Everything is negotiable," Gray told him.

McCarter was willing to try. Uncertain that the gambit would work, however, he enlisted his friend as an unpaid consultant. He gave Gray material about Sue and asked him to try to figure out how much she would bring if she did go to auction. "John's people had done a financial analysis," Gray says. "They had tried to rationalize what kind of value you can get, even down to souvenir sales."

Williams and Sotheby's must also have rationalized Sue's value. They turned down McCarter's preemptive offer, which was in the neighborhood of $1.5 million. Having read more about the dinosaur by this time, Gray understood why. "I got a growing sense of her iconic value and how various parties around the globe might react," he says. He reasoned that if McDonald's and Disney were on board, "there would be serious competition and not just from the United States. There were all kinds of possibilities of parties who might want this as a trophy."

Could he envision casinos bidding for a *T. rex*?

"Absolutely," says Gray.

Having seen the effect of the "trophy mentality" in the art world of the late 1980s, Gray "came up with a number considerably higher than the dollars [McCarter] was working at." He suggested the museum go back to its backers or find new ones and ask for more money. McCarter and his staff sought and received additional support from certain board members and a commitment from the California State University system to provide up to $700,000 for restoration and study of Sue. Putting this auction package together was "a bit like space station contracting," says Flynn. "You don't know what the upper limit is since there is no precedent." Fearful that word of the museum's interest in Sue would attract other bidders and drive the price up, Gray insisted that McCarter tell his colleagues and everyone at McDonald's and Disney not to discuss the dinosaur with anyone. "Cut off all contact with Sotheby's, too," Gray ordered. "Throw cold water on it."

At one point Sotheby's called for the museum's financials—documents necessary to facilitate bidding. "Don't respond," Gray said. "Tell them, 'We're not planning to participate,' anything to turn off their interest."

Shortly before the auction, Sotheby's again called. Did the museum need help with any arrangements? Gray suspects the house was more interested in knowing if McCarter would be bidding. Again, the museum's director played dumb.

About this time, McCarter again called Gray. "Up to this point he hadn't said boo to me about [attending] the auction," says the art dealer.

"Dick," McCarter asked, "would you be willing to be one of our agents and handle this?"

A "flattered" Gray said yes. Then he called Sotheby's president Brooks, with whom he dealt regularly. "I want a room to bid from," he said.

Brooks said she would make the arrangements. She also agreed to personally take Gray's bids, which he would make from a phone in the room.

"I never told Dede whom I represented, and she never asked," says Gray. "She might have said, 'Mmm. Chicago. Must be the Field Museum.' But she couldn't be certain." To maintain that level of uncertainty, Gray arranged to bid from his own house account.

On the eve of the auction, Gray, McCarter, and Dr. Peter Crane, the museum's head of academic affairs, flew into New York. Uncertain if he had enough money in his war chest, McCarter continued to call prospective donors. At dinner that night, the three men discussed their strategy: Keep a low profile and don't do anything to contribute to a bidding frenzy that might drive up the price. They also discussed what would happen if they were successful. "I said I should make a statement," Gray remembers. "So I wrote out a few words."

The following morning they took a cab from their hotel to Sotheby's, which sits on York Avenue near 72 Street in midtown Manhattan. The auction was scheduled to begin at 10:15 AM. Their plan was to slip unseen into the auction house after everyone else had taken their places on the floor. Arriving a few minutes early, they waited inside the cab until about 10:12. A Sotheby's staffer then took them up a stairway and through an exhibit gallery to their room. Unbeknownst to them, Williams was already in the room directly across from theirs.

Sotheby's second level features a series of relatively small glass-enclosed "sky boxes" for bidders who wish to remain anonymous. The rooms are stocked with champagne and hors d'oeuvres. More important to Gray, they have curtains. "If you're skillful, you can see what's happening on the floor, but they can't see you," he explains.

Crane was assigned the task of standing by the curtains and reporting whom he recognized from the museum world and who was making each bid. Gray would do the bidding itself. And McCarter would follow along on a single sheet of paper that had three columns, marked "Bid," "Premium," and "Total." The bid plus the premium (Sotheby's 10 percent commission over and above the final bid) equaled the total. "I wanted to make sure we were focused on the right column—total," says McCarter. He feared that in the excitement of the moment, they might forget about the commission and bid more than they had to spend—a figure the museum has never revealed.

As Redden approached the podium, Gray established phone contact with Brooks. At about the same time, Peter Larson established phone contact with the Sotheby's representative who would keep him informed of the bidding. Larson had been released from prison about six weeks earlier, after serving 18 months of his two-year sentence. He had received the maximum 90 days off for good behavior but was under "home confinement" until mid-November. He was allowed to travel the 15 yards from his trailer to the institute, but he had to check in with his parole officer when he arrived at work. He could also go to the store and run other errands, but again, he had to advise the officer of his whereabouts. Under any circumstances, he had to be home for good by 9:00 PM.

Larson had returned to Hill City a changed man. "I'm certainly sorry for the pain and suffering of my family and friends, but I think that so many valuable things came of this," he said later. "I learned so much about myself. . . . I learned that love and family and friends are more important than work. It really changed my attitude toward life."

"The worst part was actually that period before incarceration when you're facing the unknown and have these horrible fears. Once you actually get to prison, you know here you are for this period of time and you'll be doing these things, so you get into a pattern. It's a lot harder on the family." He credits Kristin, who made the long drive to visit him 24 times, with helping him to survive.

Before leaving for Colorado, Larson had talked to others who had been in prison. "Everyone told me, 'Do your own time. Mind your own business,'" he says. He did just that, although he became something of a celebrity because newspapers and television stations occasionally came to interview him. Fellow inmates couldn't believe he had received two years for his sins. "They told me, 'Man, that's the worst we ever heard of,'" he remembers.

Larson's "day jobs" included working at landscaping, in the library, and as a busboy. He devoted the rest of his time to correspondence and business. In 18 months he answered more than 1900 letters of support from people in 38 states and five countries. He also wrote three scientific papers on dinosaurs and finished all but one chapter of the first draft of his book on the science of *Tyrannosaurus rex*.

While still in prison, Larson had read a newspaper report that Sue was in New York, being prepared for auction. "I said, 'Oh god, what are they gonna do to that fossil? They could ruin it or destroy it,'" he remembers. Ever the paleontologist, he sent Redden a long letter explaining "not only the methods that needed to be used [in preparation], but the fact that one of the real values of Sue is the contextual information that needs to be recorded as the matrix is removed and the little fossils that are included there are revealed. . . . I said we'd help in any way possible. It's like helping the FBI wrap her to take her away. You want to make sure that nothing harmful happens."

Redden assured Larson that Sue was being treated well. Sotheby's had retained Henry Galiano, formerly a curatorial assistant and collections manager at the American Museum of Natural History, as its paleontology consultant. The institute did help in several ways, providing information about Sue and photographs for the auction catalog. "They were extremely generous in dealing with us," says Redden. "They were the finders. To them will always go the glory of finding an extraordinary fossil. They cared about the fossil deeply."

So, too, did Sotheby's. Aware that Sue could be lost to science and perhaps the public if a private collector prevailed, the auction house had devised special sales terms for museums and educational institutions: payment over three years at no additional interest.

While Sotheby's prepared Sue for sale, Larson's wife and mother-in-law tried to find a sponsor to help get her back. Stan Adelstein, a successful contractor, eventually consented. "Unfortunately there wasn't enough money in the state of South Dakota to get her at the auction," says Larson.

Adelstein was out of the hunt before the Field Museum even entered. Why did Gray wait until the bidding had reached $5.2 million before opening his mouth and the museum's checkbook? "The last thing I ever want to do is pick up sales," he explains. Apparently he was the only professional among those bidding on Sue. "In the field of dinosaurs, there's not a lot of auction experience. In a room full of amateurs, you can't get caught up in the excitement." These amateurs could be "hyped up" and moved to bid unreasonably, says Gray. "My job was to keep cool and calm. Dede kept asking, 'Are you gonna bid?' I said, 'Maybe.'"

"Five million nine hundred thousand."

McCarter followed along on his cheat sheet, biting his nails. "I was not confident," he says. "I didn't know if we had enough money to prevail."

Crane watched the floor. It appeared that the North Carolina State Museum of Natural History and one other unknown bidder were still in the running. Gray was confident. "At a certain point, I could tell by the pace and the size of the bids that they were reaching their limits," he says. He sounds like a jockey ready to make his move when the front runners run out of steam.

> "Seven million." Pause. "Seven million one hundred thousand."
> Pause. "Seven million two hundred thousand."
> At $7.5 million, the crowd gasped and giggled nervously. Redden searched the room for another bidder, beseeched the room for another bidder. "Seven million five hundred thousand," he repeated.
> Gray smiled, more confident than ever. He spoke to Brooks. She nodded to Redden. "Seven million six hundred thousand," he said. He sounded faintly amused. "Seven million six hundred thousand. Seven million six hundred thousand. Fair warning, seven million six hundred thousand up here."
> "Put the goddamn gavel down," Gray shouted in the room.
> Redden brought down the gavel. "Seven million six hundred thousand." The room exploded into applause.

"What am I going to tell them?" Brooks asked Gray. She still did not know whom he represented.

"Tell them someone will be down."

As Gray headed for the doorway, McCarter picked up the phone. Flynn and other staffers and Block and other trustees had gathered in McCarter's office to await the results. So, too, had Amy Murray of McDonald's. "Chicago, we have a dinosaur!" announced the museum president.

Including Sotheby's commission, the dinosaur cost $8.36 million. Gray says Sue went for almost exactly what he expected. Says Crane: "We had a limit and we were getting pretty close to it."

When McCarter called, "the room erupted," recalls Murray. "There was so much emotion—shouting, jumping, tears."

By this time Gray had reached the press area at Sotheby's. The trio from the museum had realized the true gravity of the event when they first arrived in their room and saw a horde of television news crews on the floor. Now Gray faced the largest media assembly he had ever seen in an auction house. Who had won Sue? they and everyone else who had come to the auction wanted to know.

Gray pulled out the statement he had written the previous evening. "This morning I have the pleasure of having been awarded custody of Sue, the world's largest and probably oldest young lady," he said. "She will spend her next birthday—that's her 70 millionth—in her new home on the shores of Lake Michigan. That is, of course, in Chicago at the renowned Field Museum of Natural History."

Tremendous applause followed. The fears that a private collector had prevailed and would be spiriting Sue away to a castle or pagoda or casino lobby evaporated. As Flynn would later say, "'Now [Sue] can be around for another 65 million years for everyone to see."

As it turned out, Sue probably would have been available for everyone to see even if the Field had not prevailed. The North Carolina State Museum had been the second runner-up, dropping out at $7.2 million. The underbidder, who had been willing to pay $7.5 million, was the Jay I. Kislak Foundation, an entity created by a wealthy Miami financier. Kislak, a collector whose primary interest was pre-Columbian artifacts, had been taken with the story of the *T. rex*. If victorious, he planned to donate the fossil to a natural history museum in Florida.

Redden was not surprised by the outcome. He says that only one of the bidders had been foreign, then adds, "One of the issues in the media was that surely this will be bought by a private individual who would put it in a vault in Hong Kong or Asia or somewhere else. I found that unlikely. To me it was an obvious museum object. Nor would anyone have any more money than an American museum. People say American museums are poor. I think they are rich [when it comes to acquisitions like this]. They may have a tough time finding someone to fix the roof, but for a fabulous acquisition . . . ." He says that all the museums bidding had corporate support. Some had asked him to recommend prospective corporate sponsors, and, he says, "I recommended companies like [McDonald's and Disney]."

After Gray finished his statement, McCarter met the press to talk about his fabulous acquisition. He announced that McDonald's and Disney had helped the museum acquire the dinosaur, as had the California State University system. As he spoke, the fax machines at the museum were already humming, sending out a press release written with the help of McDonald's. Media outlets across the country quickly learned that McDonald's would take two casts of Sue on tour after the museum unveiled her original, reconstructed skeleton in the year 2000 and that another cast would stand in Disney's DinoLand USA.

Word of the Field's unique marriage with the corporate icons generated a buzz throughout Sotheby's. "It was kind of a dazzling moment," says Daly. "There was tremendous enthusiasm for the fact that the Field Museum had gotten it. . . . And then when John immediately announced it had been done in a consortium in a way that was unheard of, I think, in their ranks, and then that the consortium included McDonald's and Disney, I think that was even a little more dazzling. So that intuition that we all had that this could be big kind of came true that day. It was an international story. Global. It really was big."

Bigger than Peter Larson had even dared imagine. "I think we just underestimated her value," he said shortly after the auction. "I mean she's the most famous fossil in the world."

How did he feel about the auction's result? "Of course, we're sad we don't have her, but she's going to have a good home," he said. "The Field Museum is a wonderful home. The people there will treat her with respect as well as care. We're going to be able to learn again from her. She's really a time capsule. She's not just a pile of bones."

J. Keith Rigby, Jr., a paleontologist who hunted dinosaurs in Montana, viewed the result differently. "Everything changed on that day," he said at the SVP convention a week after the auction. "This sale may be the single most damaging action in the history of vertebrate paleontology."

13

YOU MAY APPROACH
HER MAJESTY

The caller from Montana had bad news. The Waltons were trying to steal Keith Rigby's dinosaur. They had moved a backhoe onto the land, and now they were trying to take the bones. Rigby, who was in Boston at the time, made some telephone calls. Local paleontologists rushed to the site. Someone from the digging party brandished a gun. Finally the local sheriff arrived to cool tempers and sort things out.

It was fairly complicated. Rigby looks for fossils around the world. As he is an assistant professor at Notre Dame, he has limited funds for his forays. But he has found a creative way to pay his way. He leads fossil collecting expeditions for Earthwatch, a Boston-based, not-for-profit organization. Amateur dinosaur hunters pay Earthwatch for the experience of accompanying and helping Rigby on his digs. During the summer of 1997, those digs were near Fort Peck, Montana—the area in which Larson was alleged to have collected off federal lands.

Rigby had received permission from one branch of the Walton family to search a promising formation. His finds would not be sold. Rather, he was working with local residents to create a natural history museum and interpretive center in the area. He believed, as Peter Larson believed, that a museum featuring dinosaurs would attract visitors and boost the economy.

In July 1997, four members of the Earthwatch party literally stumbled over some badly weathered bones. They started digging and found more remains of a meat-eating dinosaur with a jaw some 5 feet long. "If it's a kind of tyrannosaur, it's probably the largest one ever found," Rigby told the press shortly after the discovery. "It may be another type of carnivore, but whatever it is, it's an enormously important discovery."

Rigby, who employed members of the Walton family to help run the camp, decided to postpone full-scale excavation until the following year. His crew wrapped the exposed bones in plaster jackets and reburied them. Some days later he received the call that the Waltons were trying to dig up the skull of what had been named the "Peck Rex."

When confronted by authorities, the Waltons claimed the bones were theirs. "We've been around a long time," Don Walton later told the media. "One of our ancestors was a signer of the Declaration of Independence. How would you like it if the government came in and told you what you could do or not do with your land?"

Unfortunately for the Waltons, the land wasn't exactly theirs. Some years earlier a Department of Agriculture agency had foreclosed on it. Rigby knew of the foreclosure, but had been told that the plot where the dinosaur had been found still belonged to the Waltons.

In the past, like the Larsons, Rigby had taken landowners at their word. Arguably, had those landowners been mistaken or intentionally misled him, he, like the Larsons, might have been subject to prosecution if he had unwittingly excavated on what turned out to be public lands. In this particular case, however, he had grown suspicious of the Waltons after the dig began. He therefore conducted a title search. This search revealed a muddled chain of ownership. To be on the safe side, Rigby had sought a permit from the federal government shortly before the Waltons struck.

Walton insisted that his family had paid back its federal loan but had not been credited with the repayment. Thus, he said, he had a legitimate claim to the land and the dinosaur. "We need food for our families and fuel for our machines," he said. "Those people at Earthwatch say it's a nonprofit outfit, but Rigby and those people earn money to pay their mortgages and put food on their tables, don't they? Is it fair that the landowner gets nothing?"

Two weeks before Sue was to be auctioned off in New York, Rigby hurried back to Montana to finish excavating his tyrannosaur. By this time the government had determined that the land in question was indeed federal property. The Waltons had no claim to the site or the dinosaur.

"There's no question in my mind that Sue propelled the theft of our specimen," Rigby says. "[The thieves'] plan was to allow us to work and get excavation and then pull a Maurice Williams. And this was before the auction, when they thought Sue would go for a million. Supposedly they already had ours sold."

There was a sad irony here. The federal government apparently seized Sue and prosecuted the Larsons to send a message that illegal fossil collecting would not be tolerated. That seizure made Sue a celebrity who went not "for a million" but rather for $8.36 million. As a result, said Rigby, the sale of Sue, "as well intentioned as it was," may have inspired a new generation of fossil thieves. The paleontologist also expressed fears that "every rancher with fossils on his land from now on is going to demand thousands of dollars from any researcher just to look around."

Rigby may have been overstating the case, but he could point to a historical precedent. In 1923, the famed fossil hunter Roy Chapman Andrews of the American Museum of Natural History made an exciting find in the Flaming Cliffs of the Gobi Desert in Mongolia: dinosaur eggs. When he returned to America, he was deluged with offers for not only the eggs but exclusive photos of them. He resisted such overtures, but he did use the publicity to raise monies for future expeditions to the desert. To gain attention and support he auctioned off one of the eggs. A relative of one of the members of Chapman's collecting party won the specimen with a bid of $5000 and promptly donated it to Colgate University.

Wilford writes that Andrews came to regret the auction. "Chinese and Mongol officials, hearing of the purchase got the idea that dinosaur eggs were worth $5000 each on the world market. . . . 'They never could be made to understand,' Andrews said, 'that that was a purely fictitious price, based on carefully prepared publicity; that actually the eggs had no commercial value.' Future expeditions were watched with increasingly suspicious eyes."

Sue, of course, did have commercial value. But it remained to be seen whether her price was an aberration due to "all those stories." The price commanded by Nuss's and Detrich's Z. rex might be a better predictor of the future of vertebrate paleontology.

Whatever the implications of the auction turned out to be, no one in the scientific community faulted the Field Museum for acquiring Sue. Bakker spoke for his peers when he credited the institution and its corporate partners with saving the dinosaur. The museum was, he said, "a class act."

Because McDonald's and Disney had helped purchase the dinosaur, some of these same scientists and others in the general community, as well, were anxious to see if the inevitable marketing of the *T. rex* would be handled with class. Had the museum sold its soul to gain Sue? Would the scientific specimen be turned into a Happy Meal character or cartoon star?

The museum and its partners were well aware that their conduct would be scrutinized. "It's important to emphasize that by contributing, McDonald's and Disney are supporting our programs, not that they are buying a piece of anything," Flynn said shortly after the auction.

McDonald's Daly confesses, "There was some friendly push back from people working the angle that this was a potentially onerous thing because this wonderful priceless thing could be wrapped up in some kind of marketing." As a result, he says, the days following the acquisition were "an odd period for us, because we knew that the one thing everybody thought we'd do (like market a Sue Happy Meal or *T. rex* burger), we were not going to do. . . . The components of what we were going to do were in place (and announced) at the time of the auction—phase one: the fossil prep lab, phase two: educational materials, and phase three: the tour." (However, after the auction the operator of the McDonald's restaurant in the museum did introduce a dinosaur theme—shades of the Crystal Palace 144 years earlier.)

McDonald's was confident that it could meet its timetables for each of its three phases. While the fossil lab was built, the company started developing the extensive educational materials it planned to distribute free of charge to each of the more than 60,000 elementary schools in America by the coming spring. The fast food giant hoped the museum could move equally quickly in providing it the scientific information needed for these materials. Says Murray, "We're accustomed to working in drive-thru time, while museums move in geological time."

The museum knew it would have to move at a faster pace than normal—to meet not its partners' timetables, but its own. It had not publicly announced a date for the unveiling, but it was targeting May 2000. In the museum world, this was "drive-thru time." Sue's bones needed to prepared. They needed to be studied. They needed to be cast. And they needed to be mounted.

Since any given bone couldn't be in more than one place at any one time, the logistics alone were daunting. "Most objects we study we have had for a long time," says Amy Louis, a business consultant who was hired by the museum as "Sue project coordinator" after the auction. "But here we had the concurrence of science and exhibition. That's not usual. There's a challenge in doing both at the same time."

The challenge began as soon as McCarter arrived home with Sue, or at least part of her. Before leaving Sotheby's the museum director had packed up a couple of the dinosaur's incisors. He made it as far as an air-port checkpoint, where an x-ray machine picked up what appeared to be a pair of 8-inch weapons in the box he was carrying. After baring the teeth, he was allowed to proceed.

"Ladies and gentleman, I now present you Sue," McCarter told a planning meeting when he returned. In came a museum staffer carrying a towel. In ceremonial fashion, the towel was unwrapped to reveal a 67-million-year-old *T. rex* tooth. The assembled were so bitten with the dinosaur bug that Pat Kremer, the museum's director of public rela-tions, suggested the teeth be displayed to the public immediately. Spontaneous idea number one.

Spontaneous idea number two quickly followed—along with the rest of the dinosaur's bones. The positive response to the dental display inspired the museum to mount a two-month-long "Sue Uncrated" exhibit after the remainder of the *T. rex* arrived in a specially equipped moving van—complete with a security guard who stayed up all night to protect against hijackers. The skull, replastered by Simpson for the cross-country trip, was displayed, as were other bones in various stages of preparation. Visitors could also watch a brief videotape about Sue and view photos of the dig. These same visitors could also watch a preparator clean Sue's bones.

The Black Hills Institute supplied many of the photos for this exhibit. Peter Larson had also talked to Simpson about Sue. "There's a lot of difficult preparation," he said. "She prepares beautifully—the bone is just wonderful bone—but you need to know the tricks."

The paleontologist didn't expect to be asked to prepare bones. He did hope, however, that he could continue doing his research on Sue. "We'd also like to have a cast of Sue here for the people that supported us. I want to give them that at least," he said.

Larson's research would have to wait. Shortly after the auction, the museum announced that it would be hiring a postdoctoral researcher to study Sue and write what Crane called, "the definitive description—a monograph on *T. rex* based on a complete specimen." Until the monograph was completed—a task that could take up to two years—it was unlikely that anyone outside the museum would be granted extensive access, said Crane, who has since left the museum to take a post at the Royal Kew Gardens in London.

Casts the size of Sue could cost $50,000 to $100,000. Larson, however, appeared to have something he could offer the museum in return—the rights to the name Sue. The institute had tried to trademark the name shortly after the discovery. It appeared, however, that it had failed to file the proper applications. Nevertheless, Larson cited court decisions granting unregistered trademarks to companies or organizations that proved they had used a name in the past in the marketing of products.

In early December, Larson went to Chicago to consult about the preparation of Sue and to work out a deal for her name. At that time he upped his asking price to two casts and financial compensation. He acknowledged that some might consider him a gold digger, but he explained: "There's a difference between science and business. It's sort of a difficult thing for me because I'm doing both. Of course the science (with respect to Sue) is free; I'm willing to share what I know for nothing. But we've spent $209,000 on her preparation, and we spent hundreds of thousands of dollars in court. We're still paying our legal fees. I spent two years in jail. We created Sue's persona. You could always give it another name, but it wouldn't be the same." Despite the uncertainty about the Larsons' rights, the museum decided to avoid litigation and strike a deal. Negotiations continued until late January, when the museum made its final offer: $150,000 and one cast of the dinosaur.

Larson had retained a lawyer specializing in intellectual property law. The lawyer, who was to be paid a percentage of what Larson received, persuaded the paleontologist to reject the offer. Larson explained that he would have been willing to give the museum the rights for little or no money, just a cast. However, his lawyer had persuaded him that the name was worth more to McDonald's and Disney. Those companies were never party to the negotiations, as the museum owned Sue.

Once again a dispute over Sue had devolved into bitterness and misunderstanding. Larson said he thought the museum was unnecessarily playing hardball, intimating that if he didn't accept the offer, he might lose consulting contracts with some of the museum's partners. "I don't like to be threatened," he explained.

The acrimony was mutual. In a letter rejecting the institute's final offer, museum counsel Felisia Wesson wrote Larson's lawyer: "Courts have repeatedly found that your client is not entitled to any such interest [in the fossil] and we view Black Hills' current assertion of rights in the name 'Sue' as nothing more than a further attempt to profit from its original unlawful acts." (Larson notes that he was never charged with any "unlawful acts" regarding Sue.)

On January 23, the museum announced that it would discontinue using the name Sue and launch a nationwide "Name the T. Rex" contest for schoolchildren. The museum's press release expressed concern "that continuing discussions and lack of resolution would take time and attention away from important research and educational objectives."

Those research and educational objectives were already under way. The museum had hired 30-year-old PhD Christopher Brochu to study the *T. rex* and write the monograph. Brochu, whose specialty up to this point had been crocodiles, began on February 1. At the same time, McDonald's was preparing its educational materials—a children's book, a videotape, and lesson plans. Committed to a spring delivery, the company decided to call the *T. rex* Colossal Fossil instead of Sue.

The dispute ended relatively quickly, thanks to Hendrickson and Bakker, both of whom were shocked to learn of Larson's demands. They felt, as did many of their colleagues in the paleontology community, that Larson should be handsomely compensated for his past efforts. In their eyes, however, the museum's offer of money and, more important, a cast seemed fair. "It was totally out of character for Pete to demand a royalty," says Hendrickson. She and Bakker talked to their friend and then tried to broker a compromise with the museum. Tired of fighting, Larson gave up all rights to the name without demanding anything in return; he merely asked that the museum consider giving the institute a cast.

By this time the contest was well under way. The museum decided to continue to accept entries and honor its promise to award prizes for the winning names. After announcing that it had selected Dakota as the best of the more than 6000 submissions, the museum said that it had decided to keep the name Sue. The reason? Calling the dinosaur Dakota might infringe on trademarks already held by others.

A bone by any other name remains a bone. The *T. rex*, whether called Sue or Dakota, still needed to be prepared so that it could be studied and mounted. Because the museum's staff preparators were busy on other projects, Simpson hired two new teams to work on the specimen—one at the museum and one at the fossil lab in Disney's DinoLand USA. He was particularly pleased to re-engage Bob Masek, a skilled technician who had left the museum some years earlier to work with famed University of Chicago paleontologist Paul Sereno.

Preparator Simpson and researcher Brochu were hardly the twentieth-century equivalents of Cope and Marsh, though there was a certain competition for particular bones at the outset. Brochu could fully describe and study only prepared bones. Given a choice, he would have liked to start with Sue's skull. He was already making plans for the long-delayed CT scan. Simpson, on the other hand, didn't want his crew to cut their teeth on the most valuable part of the dinosaur. "We wanted to get up to speed. We had a learning curve," he explains. "So instead of starting on the skull, we picked a partial vertebra from the chest region that was not quite as preserved as other bones." Fortunately, because the institute had prepared several major bones before the raid—including the femur—Brochu had plenty to do.

The institute had found about 270 of Sue's bones. Once the preparators were up to speed, they could, for the most part, work on any of these in the order Brochu desired—if they could find them. The unprepared bones remained in their plaster field jackets. The institute had marked these jackets to indicate what was inside. Unfortunately, so, too, had the FBI and Sotheby's, and the markings weren't always consistent.

"Usually you can tell from the shape of the jacket," says Simpson. "But here there were lots of small fragments and the markings on those were hard to figure out." On opening some of the jackets, the preparators found bones they didn't think had been unearthed by the institute. In the end, they determined that only a foot, an arm, and a few ribs and vertebrae were missing.

Brochu's needs were not only subject to the learning curve of the preparators. Phil Fraley, who had been hired by the museum to mount the skeleton, was also hungry for bones—preferably the hips first. He wanted to start his model in the middle and work outward toward the tail and skull.

The New Jersey-based Fraley had established his reputation by remounting the Barnum Brown *T. rex* and several hundred other vertebrate fossils at the American Museum of Natural History. Edwin Cope, it turned out, was not the only expert ever to err in constructing a dinosaur. At the turn of the century, the American Museum's Henry Fairfield Osborn stood the Barnum Brown *T. rex* upright, its legs measuring fifteen feet from hip to floor. The decision to do so was done in part for show, in part because the technology of the day made it difficult to brace the dinosaur in a less erect fashion, and in part because Osborn did not have an entire specimen with which to work, writes Horner. In fact, "To make the *T. rex* look whole and steady, Osborn made other creative decisions. He added several feet of imaginary lizard like tail, which also helped to stabilize the stand-up *T. rex*." As a result, says Horner, "For generations, in its most famous and influential representation, *T. rex* has been viewed as a stiff, long-tailed lumbering beast."

Charles Knight, who worked at the American Museum of Natural History before painting the murals in the Field Museum, portrayed many of his *T. rex* as upright, too. The tall, plodding creatures in movies like *Godzilla* were based on one of Knight's paintings, as was a 25-foot-long animated *T. rex* at the Sinclair Refining Company's dinosaur exhibit at the 1933 World's Fair in Chicago.

In 1992, the American Museum finally dismantled its famous *rex*. After much internal debate, the staff chose a different pose: head down, back horizontal, jaws shut. Fraley's remounted skeleton was unveiled in 1994.

Prior to its 2000 unveiling, the Field Museum was not about to reveal Sue's exact pose. Flynn would say only that the museum's own experts had designed a "scientifically accurate biology of the living animal." He explained that such a design is influenced by the study of "joint surfaces that limit the range of motion" and the examination of the hind limbs and tail. "Once you set constraints, you have aesthetic latitude," he explains.

Many dinosaurs that stand in museums around the world are plaster or fiberglass casts of the original. Not Sue. Only her skull would be cast for the Field's exhibit. The museum estimates that Sue weighed 7 tons. At about 600 pounds, the real skull could not be supported by the fleshless skeleton. It was given its own space on a mezzanine looking down upon Sue and her faux head.

The decision to use the original Sue, not a cast, presented a challenge for Fraley and his team. They had to mount the skeleton in such a way that each bone could be removed for future study. This meant "no permanent glues, no holes drilled, and no anchors to the bone," says Simpson.

Fraley's solution: Each bone would be cradled in a hand-forged metal bracket—"like a diamond in its setting," said project coordinator Louis. The brackets would be hinged and locked, but each individual bone could be unlocked and removed for research.

Who would be conducting future research? The museum hoped to sponsor workshops and symposia to study specific elements of the skeleton and produce scientific papers. "We would like to let younger scientists do this research," said Flynn, who added, that like any other specimen in the museum, Sue would also be available for study by all qualified scholars. Such study would begin only after Brochu published his monograph.

Although Peter Larson had aroused the world's curiosity and imagination with his papers and snapshots of Sue's life, the museum was, as Louis put it, "pretty tight-lipped" about what it learned about Sue in the months leading up to her unveiling. Flynn, who between 1996 and 1999 led expeditions to Madagascar that found the oldest dinosaur bones on record, says a conscious decision was made to limit public pronouncements to May 1999 and May 2000. "We wanted to wait for peer review," he explains. "There is a great opportunity for public education here because of all the hype. Our job is to separate out the fantasy and storytelling and *Jurassic Park*. If we debunk hyperbolic speculation, that's not bad." Adds Simpson: "The skeleton is so complete that it actually constrains the stories."

The May 1999 science briefing centered on the results of the CT scan of Sue's skull at Boeing Company's Rocketdyne lab in Ventura County, California. In August 1998, the skull had been shipped to the West Coast after being foamed and packed vertically in an octangular crate customized by a structural engineer. Boeing conducted 500 hours of scans over a six-week period, or, as one Boeing technician put it, "more radiation than Godzilla received when the French A-bomb was detonated in Polynesia." Brochu and Masek were present for much of the work. When Brochu went back to Chicago, the Boeing operator sent him data over the Internet.

The museum received a total of 748 CT images. These x-ray slices filled eight CDs and could be viewed individually or stacked to create a three-dimensional image of Sue's skull and snout. With the assistance of a computer program, Brochu could take a virtual journey through Sue's head.

Brochu was delighted with what he saw. The scan revealed that the olfactory passage—where nerves for smell passed between the nose and the brain—was huge, he said. "This thing must have smelled its way through life." Did that indicate it was a scavenger or predator? Like Larson, Brochu said he suspected that *T. rex* was an opportunistic feeder that killed some of its food and ate carcasses it found. Among the other revelations: Sue's braincase was only "the size of a quart of milk."

Brochu was not ready to pass judgment on most of Larson's snapshots or theories. "I'd love to be able to tell you a lot of things about this specimen, but I can't yet," he told *National Geographic* magazine. "Was it male or female? I don't know. Warm-blooded or cold-blooded? I'd love to tell you this dinosaur was purple with green spots. I'd love to be able to tell you the sound it made. But I can't."

Brochu did say that he had found further evidence suggesting that birds evolved from theropods. As Larson had previously noted, bones in Sue's jaws were similar to those in modern birds. And, as Larson had also observed, Sue's skull bones and vertebrae were air-filled, like those of their feathered friends.

Larson had concluded that Sue suffered a broken fibula that had healed. Brochu agreed. He was not certain, however, that Sue had been bitten by a fellow dinosaur; what Larson thought were teeth marks may just have been infected areas. The museum's researcher also suspected that Sue had died of natural causes rather than in battle. Brochu's interim findings were the focus of an exhibit premiering at the end of May 1999: "Sue: The Inside Story."

Shortly before this exhibit opened, in-house staff began making molds of the bones prepared in the museum and at DinoLand USA.

These bones were then sent to Peter May, a highly regarded cast maker based in Beamsville, Ontario. When he finished with them, May sent the molds on to Fraley, who used them to make projections pending the arrival of the bones themselves.

To coordinate these comings and goings as well as everything else related to Sue, Louis held a multidisciplinary group meeting every Monday morning. The group included representatives of the exhibits, education, geology, marketing, and public relations departments. Topics ranged from the SueCam, which afforded Internet visitors a live view of preparation, to the Sue Web site, which offered both information and merchandise. Among the items that could be purchased on line or in the museum store: several styles of Sue T-shirts and sweatshirts ranging in price from $12.95 to $35.95; Sue baseball caps priced at about $15; a Sue coffee mug at $6.95; a mini *T. rex* skull for $68; a cast of a tooth for $44; a cast of a claw for $33; and a brass Sue ornament for $14. Louis reports that sales were brisk.

Louis also served as liaison with McDonald's and Disney. The Colossal Fossil kit distributed in the spring of 1998 generated the most favorable response from teachers of any packet in McDonald's history, says Nemeth. In the fall of 1999, the company also opened a show about Sue for elementary schools, "Ronald McDonald and the Amazing Thinking Machine."

Orchestrating the tour of the two Sue casts was more complicated. "This was the area I was most concerned about," says Daly. "We didn't need to be persuaded about the logistics of moving it. I always knew that was something we could do. The concerns I had was more on the receptivity—could we create something and make it well enough financially structured so that local markets would pick it up?"

McDonald's is organized in 40 regions in the United States. Coops within the regions buy advertising, food, and paper goods together and plan promotional activities. Once Sue was acquired, McDonald's offered each of the regions the opportunity to sponsor a traveling exhibit that would include the 45-foot cast skeleton of Sue, a video, freestanding interactive exhibits, touchable casts of bones, and interactive anatomical models that would allow visitors to control the movements of a *T. rex*'s jaw, tail, neck, and forelimbs. Headquarters provided a "pro forma of what this thing was going to look like and predicted costs," but the local markets had to choose a date and find a site—a museum or educational institution. The Field Museum had to approve the venue in each market.

From the beginning McDonald's envisioned offering the tour to the U.S. markets over the three-year period following the unveiling. "Then, hopefully on the strength of the performance in the United States, we'd be able to take it international so that it had even more value to our initial investment," says Daly. Initially, the company thought Sue would stay in a region for two weeks before moving on. "But the [Field] museum, said 'whoa.' They wanted it for two months in a market and then another month to pack it all up. They have to do it right by their museums."

As the unveiling approached, the two casts were slated to visit 15 cities, beginning with Boston and Honolulu. Other sites included St. Paul; Los Angeles; Columbus, Ohio; Indianapolis; Kansas City, Missouri; Louisville; Portland, Oregon; Fresno; Toledo; Salt Lake City; Dallas; Seattle; and Milwaukee.

Disney's task was considerably easier than McDonald's. DinoLand merely had to be reconfigured to accommodate the fossil prep lab and the cast of Sue. The *T. rex* would stand in the middle of the park, outside, but under a shelter. Educational graphics also had to be created.

How do museum experts look at these efforts by McDonald's and Disney? Does taking replicas of Sue on tour or standing a replica in a theme park cross the line between science and industry? Is this type of corporate/museum partnership cause for alarm or cause for celebration? A little background is in order.

In recent years the relationships between several museums and their patrons or exhibition sponsors have raised eyebrows and questions about the division between art and commerce. In 1998, for example, the Solomon B. Guggenheim Museum in New York mounted an exhibition called the "Art of the Motorcycle." BMW, a major manufacturer of motorcycles, sponsored the show. In 1999, the Guggenheim announced that in the fall of 2000 it would present a major retrospective on the creations of Italian designer Giorgio Armani. According to *The New York TImes*, however, "What the museum did not acknowledge was that some eight months earlier, Mr. Armani had become a sizable benefactor to the

Guggenheim." How sizable? Perhaps as much as $15 million, said the *Times*.

The Guggenheim has stated that it would have mounted the Armani exhibition even if the designer hadn't contributed. Whether this is true or not, corporations are "pushing the envelope" by demanding quid pro quos, says Ed Able, president of the 3000-member American Association of Museums. Able sees a disturbing trend in which corporations have ceased creating philanthropic budgets for making contributions to not-for-profit institutions like museums. Instead, the corporations make such contributions from their marketing budgets—and the marketing departments look for something in return.

The challenge for a museum, says Able, is "to separate corporate support from content control. If a corporation exerts influence on what's in an exhibit and how it's exhibited, [it is an] absolute violation of ethical standards." He points to "Sensation," a 1999 show at the Brooklyn Museum of Art as a cautionary tale. The museum receives funding from the city of New York. New York Mayor Rudolph Giuliani tried, unsuccessfully, to close down the exhibition because he and others found some of its paintings and sculptures offensive. Able finds this attempt at control nefarious. But he is also disturbed by the control exerted by the exhibition's major financial backer, British advertising magnate and art collector Charles Saatchi.

According to a front-page article in *The New York TImes* written by David Barstow, "The director of the Brooklyn Museum of Art gave . . . Saatchi a central role in determining the artistic content of 'Sensation,' so much so that senior museum officials expressed concern that Mr. Saatchi had usurped control of the exhibition." Apparently, when museum staffers wanted to eliminate certain paintings from the show, Saatchi overruled them.

Saatchi owned the paintings and the sculptures in the exhibition. He had also pledged $160,000 in financial support to the museum—a fact the museum concealed for months, according to the *Times*. Saatchi, like other collectors, sometimes sells works after they have been exhibited. Christie's auction house, where Saatchi has sold art, also contributed $50,000 to the show. Arguably a painting may become more valuable and draw more at auction after it appears in an exhibition.

The Field Museum would leave itself open to criticism if it accepted McDonald's or Disney's support and then mounted an exhibit on hamburgers or theme parks. Criticism would also arise if there were evidence that McDonald's or Disney pressured the museum to exhibit Sue in a particular way or compromised the scientific study of the specimen. McDonald's acknowledges that it would receive some negative feedback if it turned Sue into a Happy Meal or action figure.

Because none of these scenarios appeared to have materialized, Able and others gave the museum and its sponsors high marks for the preliminary handling of Sue. Philanthropic activities, as opposed to marketing activities, present the corporation as a good citizen in the community, says Able. "I frankly admire the way in which McDonald's has gone about sending a message that it cares about the community—be it local, regional, or national," he says.

James Abruzzo, managing director of the Nonprofit Practice Group at the consulting firm A. T. Kearney agrees. "The downside [for a museum] is if you compromise your scientific mission," he says. "But the fact that you apply commercial aspects to your operation is not mutually exclusive. The slippery or oily slope is if your sponsor, say Exxon, is involved in a spill and you are an institution that says, 'Let's downplay it.'" That is not the case with Sue, he says.

That is not to say that there aren't tangible advantages to the sponsors. By aligning themselves with the Field Museum, McDonald's and Disney "coopted or purchased a credibilty that is very important," Abruzzo says. "They've involved themselves with a very popular finding, and they haven't done it by robbing the graves of Egypt." With credibility at stake, "To commercialize this would cheapen the investment," he adds. "This is brand [enhancement]. If it was marketing, they'd be selling Happy Meals."

Abruzzo notes that McDonald's and Disney aren't the only brands benefiting from the partnership. "The advertising budget for McDonald's far overwhelms the museum's advertising budget. McDonald's will spend more than the Field could ever spend, so it helps build the Field brand, too," he says.

By the fall of 1999, the museum had picked a date for the unveiling and had planned a week-long celebration to introduce Sue to the world. In

November, the Field's public relations department sent out an ambitious press kit complete with this schedule, "facts and fun" about Sue, a description of the exhibit, a discussion of the fossil's scientific importance, and the plans of the corporate sponsors. The cover of the kit featured a full-size photograph of the *T. rex*'s gaping jaws. The caption read: "She's been waiting for you for 67 million years." Opening the kit, one came face to face with a cardboard pop-out of Sue's choppers. The main press release began:

> On May 17, 2000, Sue takes the throne, presiding over her kingdom. . . . Male or female, king or queen, no one can be sure. But of one thing there is no question: Sue rules!
>
> You may approach her majesty. Walk around her—slowly. Examine the bird-like feet, the massive legs and pelvis, the surprisingly graceful tail. Stare into her bottomless eye sockets, her razor-sharp teeth and powerful jaws.
>
> This is the real thing. Not a plastic model or a plaster cast. Not a patchwork of composite bones from different specimens. These are the fossilized bones of the single largest, most complete, and best preserved *T. rex* fossil yet discovered.
>
> At a time when many museums are displaying replicas of dinosaur skeletons, the Field Museum has strengthened its commitment to authenticity. This *is* Sue.

The museum had chosen May 17 in part because it was a Wednesday—the one day of the week when there is no admission charge. "We will throw open the doors at nine o'clock and let everyone in to see her," Louis said before the big day. Even earlier, however, the formal unveiling would take place during an invitation-only ceremony that would include specially selected schoolchildren. Sue would be the main attraction in Stanley Field Hall. In taking this white-marbled territory, the *T. rex* claimed another victim. The Riggs brachiosaurus was moved to the United Airlines terminal in Chicago's O'Hare Airport.

The curtain would literally be lifted from Sue shortly after 6:00 AM. Louis explained that the media had expressed great interest in Sue's debut. The early time would allow the morning news shows to cover the event live.

While Sue would forever be the star of the exhibition, she had a formidable supporting cast—literally. On the second-floor balcony, visitors would be able to touch casts of selected bones. Nearby, they could view some of the animated CT images Brochu studied and take a virtual journey inside the *T. rex*'s head. Video clips would recap the story of Sue from discovery to arrival at the museum, and a time-lapse video would show the mounting of the skeleton. (Installation of the skeleton began two months before the unveiling. Securing the mounting's base to the museum's infrastructure was a complicated procedure.)

Around the corner from the skull and these exhibits, visitors could watch the cleaning of other fossils at the McDonald's prep lab and view additional exhibits on the science of Sue. These would include an animated video explaining how and why the views of *T. rex* continue to evolve. In addition, Sue's gastralia (belly ribs), which were still being studied and had yet to be attached, could be examined.

The museum planned a family night for Friday, May 19. The highlight? A screening of the new computer-animated movie *Dinosaurs*. Not surprisingly, the movie was produced by Disney. Bakker, who saw early footage of the movie, proclaimed the animation superb.

The celebration was to conclude over the weekend with two world premieres (in addition to the unveiling): a theater piece written about both Sue the *T. rex* and Sue Hendrickson and a musical work, the *Cretaceous Concerto,* written by Bruce Adolphe of the Chamber Music Society of Lincoln Center and performed by the Chicago Chamber Musicians.

Hendrickson would be there, although she confessed that discovering the dinosaur has been a bittersweet experience: "I'm glad she was found, but I wish someone else had found her." The seizure, the criminal investigation, the trial, and Peter Larson's sentencing had a profound impact on her. "It shattered my faith in the United States government," she said.

After living under the stars, in tents, in motels, on barges, and ships for her adult life, Hendrickson finally decided to build a home in 1997. She was so disillusioned with America that she chose a site on an isolated Honduran island. There she planned to live quietly when not diving in Egypt and the Philippines or procuring amber in Mexico and the Dominican Republic, or looking for dinosaurs in the badlands.

Hendrickson was finishing a dive in Alexandria when Hurricane Mitch struck Honduras in October 1998. She hurried home to find her new house damaged but standing—and scores of less fortunate neighbors living under her roof, or what was left of the roof. She immediately joined in efforts to save flora and fauna and contributed a considerable amount of money to the relief effort.

The house contains few of Hendrickson's discoveries. She is much more attached to the photographs of the places she has been and the people whom she has met and worked with. "I'm not a collector," she explains. "For me the reward comes in the moment of finding, not possessing."

Married once, now single, she lives with and is accompanied everywhere by her golden retriever Skywalker. Although she has never had children, she has become close with youngsters in all the places in which she has traveled and has set up university scholarships for several in the Dominican Republic and Honduras.

Hendrickson confesses that she is torn between settling down and pursuing other dreams. She has survived cervical cancer diagnosed in 1990 but has a disabling vascular problem and infection in her left leg. "I'm feeling old," she says. "But there are still a lot of things I'd like to find."

Her wish list includes a Siberian mammoth. If she finds one, she'd like to display it in a giant block of ice that museum visitors could touch. "It's hard to design an exhibit that will make someone walk into a museum and go, 'Wow!'" she says. "But if they don't, they are going to be bored and not read about it and learn. If you can stir children to want to learn anything, you're improving society. Maybe that's why Sue was calling me. So I could come get her and we could make kids say, 'Wow!'"

Another dinosaur hunter, Keith Rigby, hoped to be "in the field," not at the Field, when Sue was unveiled. He still had unfinished business in Montana. The dinosaur he found may be as complete as Sue and even larger, he said. He planned to supervise preparation and lead more expeditions.

Little has changed in the field since the auction, Rigby said. He knew of three more cases in which fossil thieves had robbed dinosaur excavation sites. "They know our license numbers. They follow us. And when we're gone, they come in and steal," he lamented. "There is now an incredible black market fueled by theft of fossils from public land. [The auction] was good for Sue, but it was the death knell for paleontology as we know it."

Vince Santucci, who investigated the Black Hills Institute in the mid-1980s, hoped to see Sue after the unveiling. The ranger was not as pessimistic as Rigby. Although he had observed that private landowners had started charging collecting fees, he thought that the prosecution of the Larsons had deterred commercial collectors from stealing from public lands, "or at least it's driven them underground," he said. He noted, however, that after a steady decline for several years, fossil crime did increase in 1999.

Bob Bakker isn't willing to sound the death knell for paleontology. If there was a black market and if private landowners were charging exorbitant collecting fees, "it's a temporary aberration," he says. Dinosaurs are still being found on private land, he notes. He points to the exciting discovery in Belle Fourche, South Dakota, of a virtually fully complete *T. rex* juvenile by a team from Houston in 1998. Recent expeditions abroad by the University of Chicago's Sereno, the Field Museum's Flynn, and Rigby himself have also unearthed remarkable specimens. Bakker also notes that Nuss and Detrich were unable to sell their Z. rex. "I heard they even put it up on eBay for a while," he says.

Bakker is less concerned about continuing to find fossils than he is about continued efforts to crack down on legitimate collectors. "We should worry about the odious behavior of small-minded PhDs and even smaller-minded government officials," he warns. He cites a recent case in which the government prosecuted a lifelong collector of prehistoric turtle fragments. The collector was charged with stealing hundreds of thousands of dollars worth of fossils. At the trial an independent expert testified that, in reality, the bones were worth less than $1000. The judge put the collector on probation instead of sending him to jail, as the government wished.

As the unveiling approached, Bakker, Peter Larson, and others were battling those they considered small-minded on another front. The BLM was considering regulations that would severely restrict commercial collectors and amateurs. In the interim, no one—not even academics—had access to certain lands. Bakker mused that maybe the BLM should be prosecuted for the destruction of fossils, since many specimens were eroding away.

Patrick Duffy shares Bakker's feelings about the government. The lawyer, who left the firm with which he was affiliated and has his own practice in Rapid City now, says, "The government was never able to create a uniform system of laws to cover [fossil collecting]. . . . It's dangerous when, if we can't make consensus in Congress, we go down the street to the J. Edgar Hoover Building and try and make consensus."

Duffy notes that the action against the institute sent some defendants and their families into therapy. Marriages broke up. One person attempted suicide. "The government spent millions of dollars and ruined lives when no fundamental right was at issue. *Lives were ruined* and other than the *T. rex*, which was not the subject of criminal charges, we are talking about shards of bone here. That's not worth happening."

Contrary to the predictions of his critics, Duffy's nemesis Kevin Schieffer has yet to seek a judgeship or run for office. Instead, he serves as CEO of the South Dakota-based Dakota, Minnesota, and Eastern Railroad. Schieffer worries about a different kind of fallout from the case: "The tragedy from the global standpoint, once you get past these two individuals [Larson and Williams], is this: When you have a conflict between science and dollars, even if it would be in the best interest of science, the dollars seem to prevail."

At least one of those two individuals, Maurice Williams, was happy to let the dollars prevail. "I didn't really care where [Sue] went," he said. "Just to the highest bidder."

Almost nine years to the day that Sue was found, Williams was back in court on *T. rex*-related business. The August 4, 1999, *Rapid City Journal* reported that a local lawyer was suing the rancher for $836,000, or 10 percent of what the fossil brought at auction. Attorney Mario Gonzales claimed that beginning in November 1990—when Williams wrote his first letter to the institute claiming ownership of the bones— he provided "advice, expertise, and assistance" to Williams "in every aspect of his efforts to recover, market, or sell the fossil." Reporter Hugh O'Gara noted that Williams was countersuing the attorney for money he claimed he loaned him to pay off a debt to the Internal Revenue Service.

When McCarter talked to Williams about buying Sue before the auction, the rancher said he thought he had more bones on his land and suggested that the Field Museum take a look. But the museum and the rancher could not agree on the terms for exploring the property. At Sotheby's, Williams also mentioned the bones to Henry Galiano, the auction's paleontology consultant, and invited him to visit.

Why? "After all, I did help him earn $7.6 million dollars," laughs Galiano.

Galiano accepted Williams's invitation. "I went there with two friends," he says. "We were just poking around and we found bones sticking out of the same producing horizon as Sue. Maurice said, 'Take 'em.'"

Galiano brought the bones back to New York. None of the scientists to whom he showed them could identify them. "We think we may have found a new genus of the dinosaur *Pachycephalosaurus* ("thick-headed lizard")," Galiano says. On a subsequent trip, Galiano found more of this dinosaur's bones, including part of the skull. "The skeleton is there," he says.

Galiano has a "handshake agreement" with Williams. "The fossil is not mine. It's Maurice's. He's the luckiest guy in the world. He wants to sell it. Here we go again, huh?"

EPILOGUE

Eight years to the day after he lost her, he would climb into his car and make the same drive he made when she was in her machine shop prison. This time there would be no reason to stop at the School of Mines. He would go to the Rapid City airport and get on a plane to Chicago so he could see her once again.

When he saw her in 1997, she was, to his surprise, "just a pile of bones." Just out of his own prison, he was not himself then, either. He knew who he was now.

In November 1998 he had gone to Washington, D.C., and installed a cast of his second favorite *T. rex*, Stan, in the dinosaur hall at the Smithsonian Museum. Two weeks later he was back in the hall. Along with Bob Bakker and Phil Currie, he was one of only a half dozen pale-ontologists invited by the museum to a gala $2500-a-plate dinner kicking off an effort to raise funds for redoing the hall. He had stood under Stan and talked about dinosaurs with some of the most important people in his nation's capital. He had felt honored to be there. He couldn't help but be struck by how strange it was that the Smithsonian—our national museum, part of the federal government—viewed him so differently than had the Justice Department. As it happened, this was the day his post-prison probation ended.

He knew she would not be a pile of bones when he saw her in all her glory. She would be Sue again. "I hope they'll let me touch her," he said before leaving. "I think we'll have another talk and see how she's doing after all this time. I'm sure I'll cry."

Shortly after the federal government seized Sue, Native American Gemma Lockhart wrote in a column for the *Rapid City Journal:*

> The intersection of scientific thought and Lakota belief can bring a strange kind of light to the controversy over Sue. From my view, that dinosaur has a strength and life of her own. So far she has been a great provider of stories and lessons.
>
> To the tribe who lost her, and to all tribes who watched and listened, something has been learned. This time the federal government has seized stone bones, next time it might be minerals or water.
>
> The Black Hills Institute and the people of Hill City stand in disbelief. They are researchers and scientists, they are law-abiding citizens and good people, and they too have lost her.
>
> The U.S. attorney is a man of good character. He stands steadfast. He knows he is in the right regarding federal law and power.
>
> In the meantime something much greater than human action and reaction is happening here. Exactly what that is may not be for us to know or say.
>
> But I have a feeling that the 65-million-year-old dinosaur is going somewhere. Her "discovery," her bone sale, and the ensuing seizure and legal battles are just part of a gift to us—her story and her tale—the things that must happen for her to find her way home somewhere.

"You're home, Sue," Peter Larson would say when he saw her. "You're finally home."

INDEX